OXFORD
UNIVERSITY PRESS

M000316879

ESSENTIAL

BIOLOGY STAGE 9

FOR CAMBRIDGE SECONDARY 1

Ann Fullick

Richard Fosbery | Editor: Lawrie Ryan

Oxford excellence for Cambridge Secondary 1

OXFORD

OXFORD

UNIVERSITY PRESS

Great Clarendon Street, Oxford, OX2 6DP, United Kingdom

Oxford University Press is a department of the University of Oxford.
It furthers the University's objective of excellence in research, scholarship,
and education by publishing worldwide. Oxford is a registered trade mark of
Oxford University Press in the UK and in certain other countries

First published by Nelson Thornes Ltd in 2013
This edition published by Oxford University Press in 2015

British Library Cataloguing in Publication Data
Data available

9780198399865

10 9 8 7 6 5 4 3 2

Printed in Great Britain by CPI Group (UK) Ltd., Croydon CR0 4YY

Acknowledgements

Cover photograph: © Getty Images/Michele Westmorland
Illustrations: Tech-Set Ltd
Page make-up: Tech-Set Ltd, Gateshead

Although we have made every effort to trace and contact all
copyright holders before publication this has not been possible in all
cases. If notified, the publisher will rectify any errors or omissions at
the earliest opportunity.

Links to third party websites are provided by Oxford in good faith
and for information only. Oxford disclaims any responsibility for
the materials contained in any third party website referenced in
this work.

Contents

Access your support website:
www.oxfordsecondary.com/9780198399865

Photo acknowledgements

Alamy: Asianet-Pakistan, p2 (bottom); Mira, p4 (left); Eitan Simanor, p20 (top); Nigel Cattlin, p22 (top right); Karen Cowled, p29 (bottom); FLPA, p42 (top right); imagebroker, p48 (photo D); Mark Hancox, p50; Ros Drinkwater, p64 (bottom); Photo Resource Hawaii, p65 (bottom right); Israel images, p70 (top); Roberto Nistri, p70 (middle); **FLPA:** Martin B Withers, p65 (top left). **Fotolia:** p4 (right); p7; p19; p27 (top and middle); p29 (top); p31 (right); p32; p33; p41; p43; p74. **Getty Images:** p46 (top right); Derek Harkness, p68 (top right). **IRRI/Gene Hettel:** p2 (right). **iStockphoto:** p2 (top); p9 (left); p12; p15; p17; p20 (bottom); p24 (bottom); p26; p31 (left and middle); p34; p35 (right); p38; p42 (bottom left); p48 (photos B, C, E and F); p52; p55; p56 (top); p58; p59 (top left); p60 (left); p64 (top); p67 (right); p68 (bottom left); p70 (bottom); p71; p72; p75; p76; p78; p79. **Natural Visions:** Peter David, p.48 (photo A) **naturepl.com:** Patricio Robles Gi, p59 (top right). **Press Association:** Nigel French/PA Archive, p54 (top left); Chris Ison/ PA Archive, p54 (bottom left); Stephen Pond/EMPICS Sport, p54 (bottom right); Aaron Favila/AP, p77; **racingphotos.com:** p54 (top right); **Science Photo Library:** J. C. Revy, ISM, p9 (right); Claude Nuridsany and Marie Perennou, p16; Dilston Physic Garden/Colin Cuthbert, p21; Newport, p27 (bottom); Dr Jeremy Burgess, p35 (left); Susumu Nishinaga, p36; Eye of Science, p44; Ria Novosti, p53; Richard Fosbery: p67 (left). **Shutterstock:** p24 (top); p46 (bottom four images); p59 (bottom left and right); p60 (right); p62; p73; **teosinte.wisc.edu:** John Doebley , p56 (bottom left).

Welcome to **Science for Cambridge Secondary 1!**
This Student book covers Stage 9 of the Biology curriculum and will help you to prepare for your Progression test and later, your Cambridge IGCSE® Sciences.

Using this book

This book covers one of the main disciplines of science, Biology, though you will find overlap between the other two subject areas, Chemistry and Physics. Each chapter starts with *Science in context!* pages. These pages put the chapter into a real-world or historical context, and provide a thought-provoking introduction to the topics. You do not need to learn or memorise the information and facts on these pages; they are given for your interest only. Key points summarise the main content of the chapter.

The chapters are divided into topics, each one on a double-page spread. Each topic starts with a list of learning outcomes.

These tell you what you should be able to do by the end of that topic.

Learning outcomes

Key terms are highlighted in **bold type** within the text and definitions are given in the glossary at the end of the book. Each topic has a list of the key terms you should understand and remember.

Key terms

Summary questions at the end of each topic allow you to assess your comprehension before you move on to the next topic.

Summary questions

Expert tips are used throughout the book to help you avoid any common errors and misconceptions.

Expert tips

IGCSE Links prompt you to think forward to what you will learn about at IGCSE level, helping you to prepare for this transition and showing you the importance of studying Stage 9 material in preparation for Cambridge IGCSE.

 *IGCSE Link...*

Topics, text or artworks marked with a rosette icon are not part of the core curriculum, so will not be tested as such in your Progression or Checkpoint test. They have been included to assist your understanding of core topics or prepare you for topics in a subsequent stage.

Practical activities are suggested throughout the book, and will help you to plan investigations, record your results, draw conclusions, use secondary sources and evaluate the data collected.

Practical activity

At the end of each chapter there is a double page of examination-style questions for you to practise your examination technique and evaluate your learning so far.

Answers to Summary questions and End of chapter questions are supplied on a separate Teacher's CD.

Student's website

The website included with this book gives you additional learning and revision resources in the form of interactive exercises, to support you through Stage 9 Biology.

Science *in context!*

Flooding means famine

Rice is one of the most important crops in the world. Millions of people rely on rice to feed themselves and their families. Plants need water and rice grows in water, so you might expect it to survive well if the fields were flooded. Sadly, this is not the case. If young rice plants are under water for more than a few days, most of them will die. The plants that survive do not produce as much rice as usual. This can cause starvation for thousands of people in an area.

Rice growing

Floods on the increase

In countries such as Pakistan and India, the number of times the land has been flooded has increased in recent years. This has had a terrible effect on the local people who have often not only lost their homes but also lost their rice harvest.

When huge areas like this are under water, the rice crop is lost before the water goes down

New types of rice

Scientists have used new breeding techniques called genetic manipulation to produce a strain of rice which is resistant to flooding. In two years they have produced a plant which can stand being submerged for up to three weeks, and still give around 80% of the normal yield. Farmers are already growing this rice in their fields and it offers hope for the future. Managing plant growth like this and in many other ways can help to provide enough food for everybody all the time.

These new rice plants can withstand flooding and produce a good crop afterwards

In this chapter you will find out about the way plants make food by photosynthesis. You will look at the ways in which plant leaves are adapted to carry out as much photosynthesis as possible. You will also discover the importance of mineral nutrition in plants and think about ways in which people can manage the way plants grow.

Key points

- Photosynthesis is the process by which plants use carbon dioxide and water, with energy from sunlight, to produce glucose, plus oxygen as a waste product.
- The word equation for photosynthesis is:

  carbon dioxide + water $\xrightarrow[\text{chlorophyll}]{\text{light energy}}$ glucose (sugar) + oxygen

- In experiments on photosynthesis it is important to use a destarched plant which has been kept in the dark for at least a week.
- You can demonstrate that light is needed for photosynthesis using land plants with a leaf covered in card or water plants giving off oxygen.
- Plants capture the light energy from the Sun using the green chemical chlorophyll contained in organelles called chloroplasts in the leaves and other green parts of the plant.
- Photosynthesis is an endothermic reaction – it takes in more energy than it gives out.
- Gases move in and out of the leaves of a plant through pores called stomata.
- The rate of photosynthesis increases as the light intensity increases as long as no other factor is limiting.
- The rate of photosynthesis increases as the temperature increases until the point at which the enzymes are denatured by the heat.
- Light levels, carbon dioxide levels and temperature all act as limiting factors on photosynthesis in a plant.
- Plants need water both for photosynthesis and to maintain the structure of the cells.
- Plants need mineral salts from the soil to make certain vital chemicals such as proteins and chlorophyll, e.g. nitrates, phosphates and potassium.
- If plants do not get the minerals they need they suffer from deficiency diseases.
- People manage the growth of plants in a number of ways including providing extra mineral salts through fertilisers and controlling the limiting factors in greenhouses.
- First hand experience and secondary sources are important for finding out about photosynthesis and the growth of plants.

Photosynthesis

Plants make glucose in a process called **photosynthesis**. 'Photo' means 'light' and 'synthesis' means 'to make' – photosynthesis is all about making glucose using light.

Practical activity   Remember photosynthesis

Before you read on, try and remember as much as you can about photosynthesis.

Work in a group, take a big sheet of paper and write down everything you can remember about photosynthesis and the ways you can find out about it. This is known as brainstorming.

Keep your piece of paper and see how much you got right as you work through this chapter.

Almost all living organisms depend on photosynthesis in one way or another. Plants make food using energy from the Sun and they turn this food into new plant material. Some animals (herbivores) eat the plants. Other animals eat the herbivores. Fungi, bacteria and other organisms feed on the dead plants and dead animals. Throughout nature living things, including people, rely on plants for their food. Plants are vital for the survival of people all over the world.

In Stage 8 you started to look at photosynthesis. In this chapter you will be studying this important process in much more detail.

People need plants – directly or indirectly photosynthesis provides us with food, clothing, building materials, medicines and even the oxygen we breathe

Plants provide food and oxygen for everything else on the Earth

The basic process

Plants use carbon dioxide from the air, water from the soil and energy from sunlight to photosynthesise. They carry out photosynthesis in the green parts of the plant, mainly the leaves. This is because they capture the energy from the Sun using the green pigment **chlorophyll**. Plants use the energy from the Sun to make new glucose molecules. Oxygen is the waste product of this reaction.

Photosynthesis can be summed up as a word equation as follows:

$$\text{carbon dioxide} + \text{water} \xrightarrow[\text{chlorophyll}]{\text{light energy}} \text{glucose (sugar)} + \text{oxygen}$$

The need for water and minerals

Plants cannot survive with just sunlight and carbon dioxide from the air. They also need water from the soil. This water is needed for photosynthesis (see the word equation above). As well as this, water keeps all the cells of the plant rigid, which supports the stems and leaves.

Plants take in water from the soil through their roots. As the roots grow they become covered in root hairs, which increase the surface area of the roots. This means they can take in more water.

Plants can make glucose by photosynthesis, but they also need proteins to make their enzymes and build their cells. They need minerals from the soil to make this possible. These minerals are taken in through the roots.

Destarching plants

You can do lots of practical work to investigate photosynthesis. These experiments often use the presence of starch in the leaves of a plant as evidence that photosynthesis has taken place. To do this you need to be sure that the plant has not got any starch left in its leaves before you start your experiments.

To **destarch** a plant in preparation for an experiment into photosynthesis you leave it completely in the dark for about a week. The plant will use up all of the starch stores in its leaves during that time. Any starch you can demonstrate with iodine solution has therefore come from new photosynthesis.

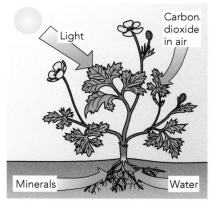

A plant needs all of these things to photosynthesise and then make proteins successfully

Expert tips

Destarching a plant before carrying out the practicals on the next few pages ensures that we can trust the results. It makes sure that we have a valid investigation.

Key terms

- **chlorophyll**
- **destarch**

Summary questions

1 Describe how photosynthesis is an important reaction for:
   a) plants
   b) all living organisms.

2 People breathe in oxygen and breathe out carbon dioxide. Some people claim that talking to houseplants makes them grow better. Can you think of a possible scientific explanation for this idea?

3 Explain what would happen to a plant if you left it in a dark cupboard for:
   a) a week
   b) several weeks.

Light is important for photosynthesis. There are lots of different ways in which you can demonstrate this fact.

Demonstrating photosynthesis

To find out about photosynthesis you need to be able to show that it has taken place. There are two main ways in which you can do this:

Practical activity — Testing for the presence of starch using iodine solution:

Some of the glucose that is made by photosynthesis is changed into starch to be stored. You can test for this starch to show that photosynthesis has taken place.

Remove the leaf to be tested and place it in boiling water for about 2 minutes to kill the cells and break down the cell membranes.

Then put the leaf into a test tube of ethanol which is placed in the beaker of hot water to heat it. This removes the green colour from the leaf.

Rinse the pale leaf in hot or cold water.

Spread out the leaf on a white tile.

Cover the leaf with **iodine solution**.

A dark blue/black colour shows the presence of starch in the leaf – which means that photosynthesis has taken place.

Testing a leaf for starch

 Wear eye protection. Do not use naked flames – ethanol is highly flammable

Practical activity — Measuring the gas given off by water plants such as *Elodea* (Canadian pondweed) or *Cabomba*

When plants photosynthesise, oxygen gas is produced as a waste product. You cannot see this in land plants. However, in water plants the oxygen produced during photosynthesis can be seen as streams of bubbles. You can count the individual bubbles or collect and measure the volume of gas given off as a way of showing that photosynthesis is taking place. By using this method, with a clock or stopwatch for timing, you can also work out the **rate** of photosynthesis.

Note that in the dark, water plants may still give off a gas but it will be carbon dioxide from respiration, not oxygen from photosynthesis.

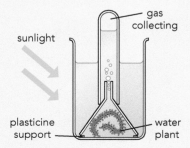

Collecting gas or counting the bubbles from photosynthesising water plants

Investigating the effect of light on photosynthesis

Practical activity   Showing that light is needed for photosynthesis

Plan ways in which you could test the idea that light is needed for photosynthesis using:

1 destarched land plants

2 destarched water plants such as *Elodea*.

Select the best possible ways of showing that light is needed. Base your ideas on what you learned in Stage 7/8. You can also carry out some extra research.

Practical activity   Showing the effect of light on growing plants

Set up a dish of fast growing seeds such as cress or mung beans on damp paper and place them in the dark. Set up another dish of the same type of seeds and leave them out in your classroom in the light. Keep both dishes damp.

After a week, compare the two sets of seeds. Put them back in their different conditions and observe them again several days later.

Your teacher may set up the seeds for you so you simply observe the plants.

Plants grown in the light make lots of new material by photosynthesis

- Describe the appearance, height, etc. of both sets of plants.

- Explain your observations. Choose some different ways to take the results.

Expert tips

When you make observations, look very carefully at all the parts of the plants: leaves, stems and root.

Key terms

- **iodine solution**

Summary questions

1 Harmesswar wants to show that a plant needs light to photosynthesise. He is given a plant and some black card. Describe an investigation he could carry out.

2 Make a table to show the advantages and disadvantages of the two main ways we have of showing that photosynthesis is taking place/has taken place in the school laboratory.

3 Darkness as well as light is important if you want to show that a plant has carried out photosynthesis. Why is this?

Chlorophyll, chloroplasts and plant cells

The green colour of the leaves and stems of plants comes from the special green chemical chlorophyll. Without chlorophyll, photosynthesis could not take place. It is the chemical that captures the energy from sunlight that plants need to make their own food.

Chloroplasts and plant cells

Chlorophyll isn't found everywhere in plant cells. It is contained in structures called **chloroplasts**. The chloroplasts contain the chlorophyll. They also have all the enzymes needed to join carbon dioxide and water together to make glucose and oxygen.
Not all plant cells contain chloroplasts. Only the parts of a plant that are both above the ground and used for making food are green.

Practical activity   The parts of plants that are not green

Work in a group. Think of as many parts of plants as you can that are not green and write them down.

HINT: think about what the different parts of the plant are for.

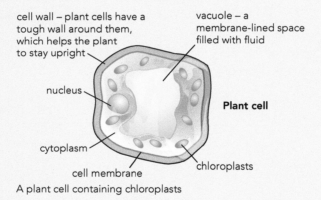

cell wall – plant cells have a tough wall around them, which helps the plant to stay upright

vacuole – a membrane-lined space filled with fluid

nucleus

Plant cell

cytoplasm

cell membrane

chloroplasts

A plant cell containing chloroplasts

Looking for chloroplasts

You are going to look at two different types of plant cells under the microscope and try to identify chloroplasts in them.

Onion cells: Take a piece of onion and remove a small piece of the very thin epidermis using forceps. Put this on a microscope slide and add a drop of water. Use a mounted needle or a very sharp pencil and lower a coverslip over your specimen. Take care not to trap air bubbles, which will look like black ringed circles under the microscope. Remove any excess water and look at the cells under the microscope.

- Draw and label what you can see.

Pondweed cells: Take a single leaf from a piece of pondweed and cut a very tiny section (about 2 mm²). Put the leaf sample onto a microscope slide and follow the procedure for putting on the cover slip described above.

- Draw and label what you can see – can you identify the chloroplasts?

Can you count how many chloroplasts are in each cell?

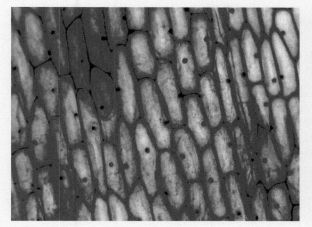

Onion cells

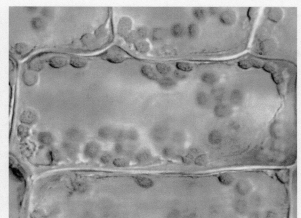

Pondweed cells

The number of chloroplasts in a plant cell depends on the position of the cells in the leaf. The cells that are near the top surface of the leaf get the most sunlight falling on them. They contain the most chloroplasts so this is where most of the photosynthesis takes place. The cells in the next layer down get less light. They have fewer chloroplasts and less photosynthesis takes place there.

A reaction that needs energy

Photosynthesis takes a lot of energy.

$$\text{carbon dioxide } + \text{ water } \xrightarrow[\text{chlorophyll}]{\text{light energy}} \text{glucose (sugar) } + \text{ oxygen}$$

All reactions need some energy to get them started. Chemical reactions also produce energy. Most reactions produce more energy than they take in. They are **exothermic reactions**. Photosynthesis takes in more energy than it produces. It is an example of an **endothermic reaction**.

Key terms

- **chloroplast**
- **endothermic reaction**
- **exothermic reaction**

Summary questions

1 Plant cells have chloroplasts and animal cells don't. Why is this statement incorrect?

2 Make an illustrated flow chart to explain how to make a slide of plant cells.

3 a) What is an endothermic reaction?

b) What is the source of the energy needed for photosynthesis?

c) How is the energy made available to the plant?

Leaves are adapted to carry out as much photosynthesis as they can.

Practical activity   Leaf adaptations

Work in a group. Collect leaves from as many local plants as you can (including trees). Draw round them all and observe them carefully.

Discuss in groups the adaptations of leaves for photosynthesis as you can by looking for the common features and remembering what you studied in Stage 7/8. Write a list of as many leaf adaptations for photosynthesis as you can. Compare your list with other groups.

Leaves and photosynthesis

Many of the adaptations of a leaf for photosynthesis can be seen on the outside. But many of them can only be seen if you look inside the leaf using a microscope. This lets you examine the arrangement of the cells.

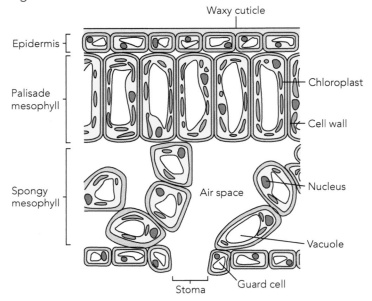

A section through a leaf. The structure of the cells in a leaf are arranged so as much photosynthesis as possible can take place

Moving gases in and out

Photosynthesis depends on gases being able to move in and out of the leaf. The plant cells use up carbon dioxide as they photosynthesise so it is important that carbon dioxide can move in from the air by diffusion. The cells make more oxygen than they can possibly use up in respiration so the extra oxygen moves out of the cells by diffusion.

The surface of the leaf is covered with a protective layer of wax. This stops the leaf from losing water all over its surface. It also stops gases moving into and out of the leaf. However, carbon dioxide can move in and oxygen moves out of a leaf through the **stomata** (singular: stoma). These are special openings in the **epidermis** of the leaf which let gases in and out. The stomata can open and close to change the amount of gas moving in and out of the plant depending on how much photosynthesis is taking place. The air spaces in the spongy layer of the leaf mean the gases can move into and out of the cells of the leaves easily by diffusion. The air spaces also allow the gases to diffuse through the leaf to all the cells. These are ways in which the leaf is well adapted for photosynthesis.

Practical activity   Investigating stomata

1 **Using a microscope:** Take a prepared slide of a section through a leaf. Look at it through a microscope. See how many of the tissues in the diagram on page 10 you can see. Look carefully in the lower epidermis for the stomata.

2 **Using nail varnish:** Paint a small part of the upper and lower surfaces of a leaf, e.g. *Tradescantia*, *Impatiens* or beans, with clear nail varnish. When it is really dry, peel off the layer of nail varnish. This layer of varnish contains an imprint of the surfaces of the leaf.

Mount the layer of varnish from the upper surface of the leaf on one slide.

Then mount the layer from the under surface of the leaf on another slide.

- Compare the numbers of stomata on the two surfaces.
- Draw a detail of what you see.

 Use a well-ventilated room. Avoid breathing in fumes from nail polish. Keep nail polish away from flames – it may be flammable.

Expert tips

Look at the leaves of plants from different habitats. Are the number of stomata on the two surfaces the same or different?

Key terms

- **epidermis**
- **stomata (singular: stoma)**

Summary questions

1 Make a 3-D model or a collage of a section through a plant leaf and label it with all the adaptations for photosynthesis that you can.

2 Why is it so important that gases can get into and out of a leaf?

3 Using a variety of sources, find out what stomata look like and draw diagrams to show what they look like when they are open and when they are closed.

The plants in tropical rainforests grow in hot, wet and light conditions. They grow big and can be spectacular. The plants of the Arctic tundra are usually small and there are hardly any trees at all. They grow in cold conditions and for much of the year they have very little light. Plants use photosynthesis to make new plant material. Perhaps there is something in the conditions that makes the rainforest plants grow so much bigger than the plants in the Arctic. You are going to investigate some of the different conditions that affect the rate at which photosynthesis takes place.

Why do you think the plants grow so differently in these two different areas of the world?

Measuring the rate of photosynthesis

When you investigate the effect of different factors on photosynthesis you need to be able to measure the rate at which photosynthesis is taking place. Using a land plant you can show simply whether photosynthesis has taken place or not using the iodine test for starch on a leaf.

By using a water plant you can measure the rate of photosynthesis. You do this either by measuring how many bubbles are given off in a minute, or the volume of gas given off in a certain time (see topic 1.3).

When you carry out an investigation you should collect data that is as **reliable** as possible. To do this, you must check that if you repeat the experiment using the same method, you get the same results. This means it is **repeatable**. You must also plan your method and record it clearly so other people can repeat the investigation and get the same results as you do. This means that your evidence is **reproducible**.

It is very important that any data you collect during your investigation is valid. You need to make sure that you control as many **variables** as possible so your measurements are **valid**. This means that you measure what you set out to measure reasonably accurately and that only one variable (the independent variable) changes.

All of these factors are very important in a scientific investigation. Remember them as you plan and carry out your investigations.

Practical activity   Investigating the effect of light intensity on the rate of photosynthesis

You are going to plan an investigation into the effect of different light intensities on the rate of photosynthesis in a water plant such as *Elodea* or *Cabomba*.

You must decide how you are going to change the light intensity for the plant. You need to do this scientifically in a way you can measure and repeat (HINT: use a metre rule).

You need to decide how you are going to measure the rate of photosynthesis from your water plant.

You need to try out your apparatus and find out if it works.

It is important to change only the variable you are investigating – in this case that is the light intensity. You must find ways to keep the temperature and the carbon dioxide levels constant throughout the investigation, for example.

- How many observations are you going to make?
- Produce a graph using your data and explain your results using your scientific knowledge and understanding of photosynthesis.

Practical activity   Investigating the effect of temperature on the rate of photosynthesis

Plan how you could adapt the experiment you carried out above to investigate the effect of temperature on the rate of photosynthesis. Remember that you must try and keep all the other variables constant.

Produce a graph using your data and explain your results using your scientific knowledge and understanding of photosynthesis.

Look at your results from these two investigations. How do you think they help to explain the differences observed between plant life in a tropical rainforest and the Arctic tundra?

Key terms

- **reliable**
- **repeatable**
- **reproducible**
- **valid**
- **variable**

Summary questions

1 Why do you need to use water plants rather than land plants to investigate the rate of photosynthesis?

2 Write clear instructions for an investigation into the effect of EITHER light intensity OR temperature on the rate of photosynthesis.

3 What other factors do you think might affect the rate of photosynthesis? Suggest how you might investigate these factors.

You have seen how light intensity and temperature affect the rate of photosynthesis in the laboratory. They also affect photosynthesis in the natural world. Levels of light and the temperature are two of the biggest factors affecting photosynthesis. As a result, they also affect plant growth – the more photosynthesis that takes place, the more the plants grow. Another factor that has a similar effect is the level of carbon dioxide in the air or water surrounding the plant. These are known as **limiting factors**.

Limiting factors

Plants need plenty of light, warmth and carbon dioxide if they are going to photosynthesise and grow as fast as possible.

- **Light** is needed for photosynthesis to take place. For most plants, the brighter the light, the faster the rate of photosynthesis. Eventually another factor limits the rate. This is often the level of carbon dioxide available to the plant.

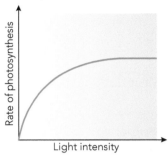

- **Temperature** affects all chemical reactions, including photosynthesis. As the temperature rises, the rate of photosynthesis increases. If the temperature gets too high it will denature the enzymes and photosynthesis will stop completely.

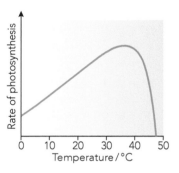

 IGCSE Link...

You will learn more about limiting factors at IGCSE. This topic is included here to assist your understanding of photosynthesis and prepare you for IGCSE. It will not be tested as such in your Progression or Checkpoint test.

- **Carbon dioxide** is needed to make glucose. The air only contains 0.04% carbon dioxide. On a sunny day the level of carbon dioxide available is often the main limiting factor for plants.

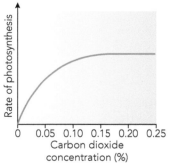

Increasing the rate of photosynthesis

We can use what we know about photosynthesis to help grow bigger and better crops. We can also use our knowledge to help us grow plants in countries where they would not usually survive. Some countries have ideal conditions for plants all year round, with lots of sunshine and warm temperatures. Even in these countries, carbon dioxide can be a limiting factor.

In many countries, or parts of countries, growing conditions are not so good. It may be very cold for part of the year (winter). The light levels may be too low or the days too short for plants to grow fast for part of the year.

Farmers and commercial plant growers build big greenhouses and grow their crops inside. A greenhouse is a building made of glass or plastic. Inside the greenhouse you can control the conditions so the plants grow as fast as possible.

There are no limiting factors for these tomatoes in the greenhouse

Key terms

- **limiting factor**

Summary questions

1 Make a big poster to explain the idea of limiting factors in plant growth.

2 Sketch a graph to show what you would expect to happen if the plant in the graph showing the effect of light intensity on the rate of photosynthesis was:
   a) given more carbon dioxide
   b) given less carbon dioxide.

3 Explain why the shape of the graph for the effect of temperature on photosynthesis is so different from the graphs showing the effect of light intensity and carbon dioxide levels.

Plants need water. If you forget to water your houseplants they wilt. If you continue to forget to water them they will die. If a country does not get rain for months, the crops will die in the fields and animals and people will be hungry.

Why is water important?

Plants need water for photosynthesis – without it they cannot make glucose.

Plants also need water to keep their cells firm so they can support the stems and leaves.

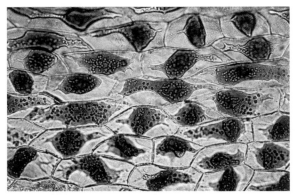

The effect of a lack of water on plant cells – they become soft and cannot help support the plant tissues

If you keep a whole plant or a single leaf in the dark, you can show that photosynthesis does not take place without light. If you deprive a plant of carbon dioxide you can demonstrate that photosynthesis doesn't take place. But if you don't give a plant water it will die! This means you cannot show that water is needed for photosynthesis in the school laboratory.

Scientists need to use radioactive water to show that water is needed for photosynthesis. They track the radioactivity and find it in the glucose in the leaves. You can find out how scientists do this by using secondary sources such as books and the internet.

Too little water

In many countries around the world there is very little rainfall. Wherever water is scarce, it is hard for plants to grow. Some plants have adaptations that allow them to survive where there is little water.

- They may have very deep, or very wide spreading roots so they can take up any water that is there in the soil.
- They may have very thick protective outer layers to prevent water loss.
- They may store water in their tissues when it rains, and then use it during the dry seasons.

Sometimes people supply the water that plants need by **irrigation**. This means that they can grow crops in dry areas. The water may be carried in ditches or by pipes and pumps. Irrigation is vital to farming in many parts of the world.

Dubai has some spectacular examples of how irrigation can enable plants to grow in a desert

Too much water

Although plants need water, they can have too much. Plant roots get water from the soil, but they also need air for the cells to respire. If the ground is flooded and becomes waterlogged, there is no air in the soil. The roots die and rot, and so the whole plant will die as well. After flooding, farmers can lose all of their crops. This has a big effect financially and it can also lead to a great lack of food.

Some plants are specially adapted to live in waterlogged conditions. For example the roots of mangrove trees are always in waterlogged soil. However, the trees have special aerial roots in the air as well, so they can get the oxygen they need to respire.

 IGCSE Link...
You will learn more about how and why plants rely so much on water in IGCSE Biology.

Key terms

- **irrigation**

Summary questions

1 **a)** You cannot show that plants need water for photosynthesis in the school laboratory. Explain why.

   b) Using other resources such as books and the internet, find out how scientists show that water is needed for photosynthesis.

2 Draw a table with two columns – 'Too much water' and 'Too little water'. Show the main problems for plants and how these problems can be overcome.

3 Using a variety of sources, make a timeline of how irrigation has been used over the years to make plants grow in dry conditions.

Learning outcomes

After this topic you should be able to:

- explain the importance of mineral salts to plant growth
- describe how to demonstrate the importance of mineral salts to plants experimentally.

The minerals that plants need are in solution in the soil water. They take them up by their roots. Plants use energy to move the mineral salts from solution in the soil water into the root cells. The minerals are transported in the xylem to cells all over the plant. You are going to find out why these mineral salts are so important to the plants and how they affect their growth.

The essential minerals

Plants make glucose by photosynthesis. This glucose is turned into starch. But plant cells need lots of other chemicals to survive. In particular they need proteins for enzymes and many other important functions in the cells. Protein molecules do not contain only carbon, hydrogen and oxygen like glucose ($C_6H_{12}O_6$). Chlorophyll contains lots of different elements too including magnesium. So plants must have elements such as **nitrogen**, **phosphorus** and **potassium** to make the chemicals they need. These are usually taken into the plant in the form of salts such as potassium nitrate and potassium dihydrogen phosphate but they are absorbed as soluble ions.

Deficiency disease in plants

If people do not get the minerals they need in their diet they suffer deficiency diseases such as anaemia. If plants do not get the minerals they need from the soil, they also suffer from mineral deficiency diseases. These cause problems in the growth of the plants and the symptoms can be seen in many different ways:

Mineral ion	Use in plant	Deficiency symptoms
nitrate	making amino acids, proteins, chlorophyll and many other compounds	the growth of the plant is stunted the older leaves turn yellow normal leaf
phosphate	making many compounds including the DNA and the cell membranes	there is poor root growth and the younger leaves turn purple normal leaf

Mineral ion	Use in plant	Deficiency symptoms
potassium	needed for the enzymes in respiration and photosynthesis to work	the leaves turn yellow with dead spots

Key terms

- **nitrate**
- **phosphate**
- **potassium**

If plants are deficient in these mineral salts, they will not grow very well. Eventually they will die.

Practical activity   Investigating the mineral needs of plants

Duckweed is a small plant that grows on the surface of water. It grows very fast so it is ideal for carrying out experiments into plant growth.

Set up five shallow dishes:

Dish A contains water with the recommended concentration of a liquid plant 'food', which gives a plant all the mineral ions it needs for growth.

Dish B contains water with half of the recommended concentration of the liquid plant food.

Dish C contains water 25% of the recommended concentration of liquid plant food.

Dish D contains water with only 10% of the recommended concentration of plant food.

Dish E contains pure distilled water with no mineral salts at all.

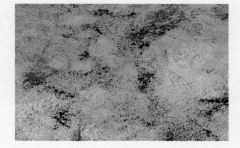

Duckweed grows very fast in its natural environment, which makes it a useful plant to observe in experiments

Add a few duckweed plants to each dish and then add a drop of oil to each dish. This will stop the water evaporating away during the investigation.

Leave the dishes on the side of the classroom for some time and make regular observations of how the plants are growing and spreading.

- After several weeks display your observations and draw your conclusions.

Summary questions

1 a) Where do plants get the mineral salts they need?

   b) Plants can photosynthesise and make their own food. Why do they need mineral salts?

2 Make a poster to summarise the requirements for plant growth, using arrows to show the intake of substances and the materials produced by the plant. Show the transport routes in the plant through the xylem and phloem as well.

For many centuries people have tried to make their food crops grow as well as possible. This often involved giving them extra water and mineral salts. Now in the 21st century, we have more ways than ever of managing the growth of our plants.

Fertilisers

As plants grow they take minerals out of the soil. In the natural world when plants die they decay and the minerals are returned to the soil for the next generation of plants to use. When we harvest crops to eat, this does not happen.

Manure returns mineral salts to the soil so the next crop grows well

Farmers can replace the minerals taken from the soil by using **fertilisers**. These are substances rich in the minerals that plants need.

We can use natural fertilisers such as animal manure or chemical fertilisers. The chemical fertilisers cost more money but contain measured amounts of minerals. Using fertilisers helps to make sure that crop plants grow as well as possible.

Computer-controlled greenhouses

Another way in which we can manage plant growth is to control the limiting factors. This ensures that photosynthesis can take place as fast as possible for as long as possible. To do this we have to tightly control the conditions in which our plants are growing. In some countries farmers do this by growing crops in huge greenhouses. Inside these, all the conditions are carefully controlled by sensors and computer systems. In the greenhouses:

- carbon dioxide levels are increased during the day so as light levels increase, the carbon dioxide level goes up
- the ideal temperature is maintained all the time
- artificial lighting is used so plants can photosynthesise for longer or to increase the light intensity.

It costs a lot of money to run big greenhouses like this. However, the plants grow as well as possible so the farmer makes a lot more money from the crops.

In some regions all the farming takes place in controlled conditions

Plants without soil

People have used fertilisers for thousands of years. This is very important around the world for the farmers who want their crops to do well and all the billions of people who rely on plants for food.

What scientists have realised in recent years is that although plants need mineral salts, they do not need soil! So now some farmers grow plants such as salad vegetables in big computer-controlled greenhouses with no soil at all. They just give the plants plenty of water containing all of the mineral salts vital for growth. This allows the plants to grow very fast and they are also very clean because there is no soil on their roots. This method of growing plants is called **hydroponics**.

Hydroponic farming is very different from traditional working in the fields

Expert tips

Some plants can grow with their roots in water so long as they get plenty of oxygen. Others need to be grown in something solid, such as blocks of rockwool (made from ground-up rock).

 IGCSE Link...
You will learn more about how farmers and growers produce food in IGCSE Biology. You will also learn about novel foods, such as mycoprotein (protein made from a fungus).

Key terms

- **fertiliser**
- **hydroponics**

Summary questions

1 a) In each of the following situations, one factor in particular is most likely to be limiting photosynthesis. In each case suggest which factor it is and explain why the rate of photosynthesis is limited:
   i)   a wheat field in a temperate country early in the morning
   ii) the same field in the middle of the day
   iii) plants growing in the lower regions of a rainforest.
   b) Explain how growing plants in a greenhouse can avoid the problems of limiting factors.

2 Explain the advantages and disadvantages of using first hand and second hand evidence to find out about the ideal conditions needed for the growth of plants.

3 Explain why farmers test soils to find out how much of the different minerals they contain. Describe what they can do to overcome any difficulties they may find.

1 Explain the difference between the following pairs of terms:
- photosynthesis and respiration
- chlorophyll and chloroplast
- palisade mesophyll and spongy mesophyll. *[6]*

2 The diagram shows a plant cell.

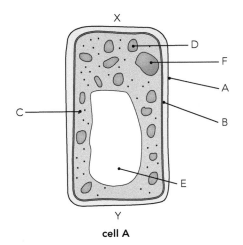

cell A

a Copy the table. Complete the table by identifying the structures in the plant cell and giving their functions. Some of the functions have been entered for you.

Structure	Letter from the diagram	Function
cell membrane	B	controls movement of substances into and out of the cell
cell wall		
chloroplast		
cytoplasm		respiration
nucleus		
vacuole		storage of water and minerals

[8]

b The actual length of the cell from X to Y is 0.1 mm. Calculate the magnification of the cell shown in the diagram. *[2]*

3 Luis and Sonya were studying the effect of light on plant growth. They germinated some seeds of white mustard in the light and in the dark. They took a photograph of the seeds:

a Describe the differences that you can see between the two groups of seedlings. *[3]*

b Explain the differences that you have described. *[3]*

c Describe three ways in which leaves are adapted for photosynthesis. *[3]*

4 a Plants require carbon dioxide and water for photosynthesis.
  i Describe briefly how plants obtain the these two substances from their environment. *[4]*
  ii Write the word equation for photosynthesis. *[6]*

b Explain why photosynthesis is an endothermic process. *[1]*

Sarah and Nabihah put some *Cabomba* into a beaker of warm water. They added some sodium hydrogencarbonate to the water and then turned on a desk lamp and directed it at the plant. They placed a funnel over the plant with a test tube filled with water on top. After a while, some gas had collected in the test tube displacing some water. Sarah measured the volume of gas that they had collected. Nabihah suggested that they use the apparatus to see what happens when light intensity is changed. When they placed the apparatus in the dark no gas was collected.

The table shows their results.

Light intensity / arbitrary units	Volume of gas collected / cm³ per 5 minutes			
	1	2	3	mean
5	3	3	4	3.3
10	6	8	7	7.0
15	3	9	11	10.0
20	12	13	12	
25	15	12	11	12.7

c Calculate the mean gas collected for the light intensity of 20 arbitrary units. [1]

d The students crossed out one of their results and did not use it. Suggest why they did this. [1]

e Plot a graph of the results. [6]

f Describe the results shown in the graph. [3]

g Explain the effect of light intensity on the volume of gas collected. [3]

5 Barley is an important cereal crop. The effects of deficiencies of the mineral elements nitrogen (N), potassium (K) and phosphorus (P) were investigated by growing barley plants with their roots in mineral solutions in tubes A to D as shown in the drawing.

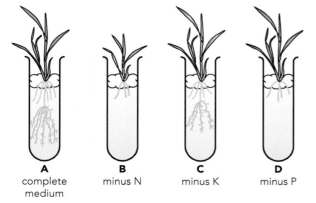

| A complete medium | B minus N | C minus K | D minus P |

Tube A contained a solution of all the mineral elements that plants require (complete medium); tube B contained all the mineral elements except N; tube C contained all except K; tube D contained all except P.

The drawings show what the plants looked like after several weeks' growth.

a Compare each of the plants in tubes B, C and D with the plant in tube A and write a description of the effects of deficiencies of the three elements (N, K and P) on the growth of the barley plants. [5]

b Explain why it was important to include tube A in the investigation. [2]

c Farmers often test their soils to see how much N, P and K they contain. Explain why it is important that they do this. [3]

Science *in context!*

Plants around the world

All around the world there is an amazing variety of plant life. Flowering plants provide us with beautiful flowers, food, fabrics, building materials, drinks, medicines and much more. The flowers are the way the plants reproduce. Fruits full of seeds are the result. Lots of these fruits and seeds are good to eat – but for the plant, they are vital to make more plants.

These mangoes each contain one very large seed

Seeds as stores

Each seed contains a tiny new plant. It also holds a store of food used by the little plant until it has grown enough to photosynthesise and make its own food. Seeds can last for many years in the right conditions.

Each seed contains everything needed for a new plant to grow

Seeds, gene banks and the future

As the human population grows and conditions around the world change, more and more types of plants are dying out. Scientists are trying very hard to save as many types of plants as they can, because we need them to help us breed the best possible crops, and to find new foods and medicines.

The best way to save plants for the future is to have seed banks, also known as gene banks. The seeds are partly dried out and then stored at around −20 °C. Most of them will last for around 200 years in these conditions. Everyone hopes that by then many of our environmental problems will have been solved. There are massive seed banks in several countries including one of the biggest in India.

In this chapter you will find out about sexual reproduction in flowering plants. You will discover the different types of flowers and the way they are pollinated. You will find out about how fruits and seeds form and the part played by the wind and by animals in dispersing the seeds.

Saving seeds in a gene bank hopefully means the plants will be there for our children and grandchildren to see

Key points

- Most flowering plants have the same structures – roots, stems, leaves and flowers.
- Flowers contain the sexual organs of the plant needed for sexual reproduction.
- The life cycle of a flowering plant always includes: germination, growth, flowering, pollination, fertilisation, seed formation and seed dispersal.
- A flower usually contains the following structures: sepals, petals, stamens made up of anthers and filaments, carpels made up of stigma, style and ovary.
- The male gametes are contained in the pollen grains made by the anthers. The female gametes are contained by the ovules found in the ovary.
- Pollination is the transfer of pollen from the anther to the stigma. Some plants are pollinated by insects and some are pollinated by the wind.
- Insect pollinated flowers are usually big, brightly coloured, scented, with nectar and small amounts of large spiky pollen grains.
- Wind pollinated flowers are usually small, green, with the anthers and stigma hanging outside the petals and large amounts of tiny, light pollen grains.
- Self-pollination is when pollen from a flower lands on the stigma of the same flower or onto the stigma of another flower on the same plant.
- Cross-pollination is when pollen from one plant reaches the stigma of a flower on a completely different plant.
- A pollen tube grows out of the pollen grain and grows down the style into the ovary. The pollen nucleus travels down the tube into an ovule and fuses with an ovule nucleus.
- The fertilised ovule grows into a seed. This contains an embryo plant and a food store for the new plant.
- Seeds need to be dispersed as far as possible from their parent plant to avoid competition for light, water and minerals with the adult plant or the many other seeds.
- Fruit can be dispersed by animals, wind, water and by self-dispersal.

Flowering plants provide us with food, clothing, building materials, fuels, drinks, medicines and more. The basic structure of all flowering plants is the same but there are many different variations around the world.

A flowering plant

The structure of flowering plants

Most flowering plants have roots, a stem, leaves and flowers. Each of the structures of the plant carries out particular functions:

The roots: The roots anchor the plant in the soil so it does not blow away. They take water and minerals from the soil and supply them to the rest of the plant through the xylem.

The stems: The stems support the leaves, flowers and fruit of the plant. They contain the transport tissues xylem and phloem. The xylem carries water and mineral ions up from the roots to the leaves, buds, flowers and fruits. The phloem carries dissolved food from the leaves where it is formed to the rest of the plant.

The leaves: The leaves are adapted to make food for the plant through photosynthesis. They have the green pigment chlorophyll which captures energy from the Sun. Chlorophyll is contained in chloroplasts which have all the enzymes needed to combine carbon dioxide and water to make glucose and oxygen. The leaves also carry out gas exchange and water evaporates from the surface through the stomata.

The flowers: The flowers contain the sexual organs of the plant. They usually appear only at certain times of the year. They may be adapted in different ways to make sure that **pollination** and **fertilisation** take place.

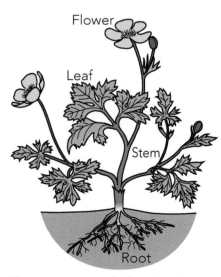

The main structures of a flowering plant

Variations on a theme

Although most plants have the same basic structure, there are many different variations around the world. For example *Rafflesia* is a group of flowering plants found in Malaysia, Borneo and Indonesia. This lives as a **parasite** on vines. It does not have any true roots, stems or leaves. The only thing you can see is the flower – and that is enormous, one of the biggest flowers in the world.

Many trees are flowering plants and they have a rather different structure with a large, woody trunk holding up the branches, stems, leaves and flowers. For example *Delonix* (or *Poinciana*) *regia* is a tree commonly seen in countries such as India and Sri Lanka. It has very bright red flowers and very finely divided leaves.

Rafflesia arnoldii – a very unusual flower, which can be up to 1 metre across

Delonix (or *Poinciana*) regia is such a striking plant it has many names. It is known as the Flamboyant tree, Royal Peacock flower, Royal Gold Mohur or the Fire tree. In Tamil it is Mayaram, in Bengali it is Krishnachura and in Marathi it is Gulmohr.

In contrast the smallest flowering plant is *Wolffia*. It is aquatic and has a single tiny leaf and no roots or stems. Very occasionally it produces tiny flowers direct from the leaf. The leaf is less than 1 mm across!

Wolffia: the smallest flowering plant in the world (showing a wasp for scale. Each plant is just two tiny leaves.

Practical activity   Investigating local plants

Go out into your local environment and find as many different plants as you can.

- Identify the plants, measure them and sketch them to show the main structural features that they have.

Key terms

- **fertilisation**
- **parasite**
- **pollination**

Summary questions

1 Draw an annotated diagram of a typical flowering plant.

2 Make a leaflet identifying five local plants and identifying their main features so people can go out and try to find them all.

3 Using secondary sources (including this text book), make a poster showing some of the different forms of flowering plants from around the world.

Without reproduction all the different types of organisms on the Earth would die out. So reproduction is very important for flowering plants. The basic life cycle is the same for all flowering plants but there are several different variations on how it works.

The main stages of the life cycle

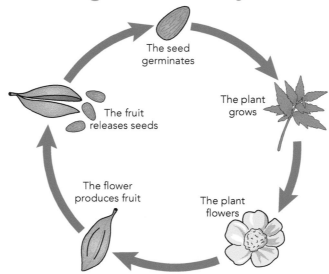

General life cycle of a plant

The life cycle of a plant starts with a seed.

- When conditions are right **germination** will start. This is the process when the tiny new plant inside the seed starts to grow. The root bursts through the coat of the seed and starts to grow down into the soil. The tiny shoot pushes up from the seed and grows into the light. The first leaves open and the new plant starts to photosynthesise.

- The plant continues to grow and develop. This stage can take anything from days to hundreds of years, as you will see.

- Once the plant is mature it will produce flowers. These contain the male gametes (sex cells) in the pollen grains and the female gametes in the ovules (see page 37).

- It is very important that the pollen is transferred from one flower to another. This is **pollination**. Once a flower has been pollinated, the male and female gametes must meet. This is **fertilisation**.

- When fertilisation has taken place, a seed forms. The plant uses food it has made during photosynthesis to build up a food store in the seed and to feed the growing embryo plant.

- The female reproductive parts of the flower also form a fruit around the seed or seeds. This is important for the **dispersal** of the seeds. At this stage, the seeds are spread as far as possible from the parent plant. Once the seeds germinate, the life cycle starts again.

The life cycle of annual and biennial plants

Annual plants complete their lifecycle within a year. Some of them can germinate, grow, flower and set seeds within three weeks! For example, *Brassica campestris* is widely used in schools because it completes its life cycle so quickly. Once an annual plant has produced seeds which have been dispersed, the adult plant dies.

A biennial plant has a life cycle which is very similar to that of an annual plant – it just takes longer. The seed germinates and the plant starts to grow. However it does not flower in its first year of life, it just gets bigger and stronger. In the second year it will flower and set seeds, and then the parent plant will die.

These annual plants produce many brightly coloured flowers but they do not last long

The life cycle of perennial plants

Not all plants germinate and complete their life cycle in a couple of years. Some plants live for years and years before they produce flowers for the first time. They begin to reproduce and do not die once they have flowered and produced fruits and seed. They continue to flower and set seeds throughout their life which may last hundreds or even thousands of years. They grow and get bigger all the time and can produce thousands if not millions of fruits and seeds during their lifetimes. These perennial plants include all the bushes and trees that you will see in your local environment.

A sacred fig tree called the Jaya Sri Maha Bodhi in Sri Lanka is known to have been planted in 288 BCE, so it is over 2300 years old

Key terms

- **dispersal**
- **fertilisation**
- **germination**
- **pollination**

Summary questions

1. Using secondary sources such as books and the internet, write a magazine article on the oldest flowering plants in the world (HINT: they are all trees). Explain their life cycles and see if you can work out how many times they have reproduced.

2. Draw a diagram of the life cycle of:
   a) an annual plant
   b) a biennial plant
   c) a perennial plant.

Flowers are part of our lives. Around the world people use them in different ways. They can be given as presents to say 'hello' or 'thank you'. They can be given to celebrate a birth or a wedding, or used to show respect to the dead.

But flowers are not just attractive to look at. They play a vital part in the reproduction of plants.

The structure of a flower

Flowers can be very different, but most of them have some features in common. Flowers contain the male and female sex organs of the plant and produce the male and female sex cells (the gametes). The male gametes in a plant are in the **pollen grains**. The female gametes are in the **ovules**.

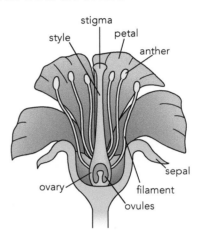

A typical flower

The different parts of the flower each have different functions:

- The **sepals** protect the flower when it is a bud.
- The **petals** protect the sex organs of the plants. They may also be used to attract insects or other animals to the plant to bring about pollination.
- The **stamen** is the male part of the flower. It is made up of the **anther** and the **filament**. The anther makes the pollen grains which contain the male gametes. The filament is the stalk that holds up the anthers that may be covered in pollen.
- The **carpel** is the female part of the flower. It contains the ovules which are made in the **ovary**. The **stigma** is the place where the pollen grains land. The **style** is the stalk that connects the stigma and the ovary so the pollen can reach the ovules.

Practical activity	Looking at flowers

Your teacher will give you a flower to look at. It should have most of the structures in the figure opposite, but all flowers are different so you may not see them all.

1 Look carefully at your flower from the outside.
- Draw and label what you can see. Measure the length and width of the flower and show this on your drawing.

2 Remove the sepals if there are any, count them and draw one.

3 Carefully remove the petals one at a time.
- Count the petals and draw one of them.

4 Remove the stamens and count them. Examine them closely – use a magnifying lens if possible. You may see pollen grains on the anther. Measure the length of the filament and the anther.
- Draw and label a stamen.

5 Examine the carpel.
- Measure the style and then draw the carpel.

6 Using a sharp knife, scissors or even a fingernail, cut through the ovary so you can see the ovules inside. Use the magnifying lens to look at them closely. Count the number of ovules if you can.

Different flowers

There will be many different flowers growing around your school and home, in gardens, in the countryside or in a park. Here are just some of the different types of flowers which can be found in the world. Have a look at them and see how many of the typical features of a flower you can see. Look at your own local flowers as well.

Flowers come in many different shapes and sizes

Summary questions

1 Name the male and female parts of a flower.

2 Collect three different flowers or use the ones in the photographs above. How many of the typical features of a flower can you see? Draw a table to compare your flowers with the typical flower in the diagram opposite.

Insect pollinated flowers need to attract pollinators such as this bee

Hibiscus flower

For sexual reproduction to take place, the male sex cell and the female sex cell have to meet and join. Flowers cannot move around to find a partner and they cannot mate. They have to rely on other ways of moving the male gametes around.

What is pollination?

Pollination is the transfer of pollen from the anthers to the stigma. This may be the pollen from one flower being transferred:

- to the stigma of another flower on a different plant, or on the same plant
- from the anther to the stigma in the same flower.

There are two main ways in which pollen is moved from one flower to another.

- In insect pollination, insects carry pollen from one flower to the next.
- In wind pollination, the pollen is blown by the wind from flower to flower.

Insect pollinated flowers

These flowers rely on insects to transfer the pollen from the anther of one flower to the stigma of another. They need to attract insects to visit the flower and they needs to make sure that the insects pick up as much pollen as possible. An insect-pollinated flower looks like the typical flower in the diagram of the flower in topic 2.3 and the flower in the photograph of a hibiscus.

Most insect pollinated flowers have special features which increase the chance that they will be successfully pollinated. These include:

- Relatively large, brightly coloured petals which attract the insects.
- Patterns on the petals which attract insects and guide them into the flower (sometimes these are coloured, sometimes they are in ultraviolet which is invisible to human beings but visible to insects).
- Some flowers produce scent to attract insects. This may be pleasant or it may smell of rotting meat or faeces to attract flies to pollinate the flowers.
- Some flowers produce **nectar**, a sweet sugary liquid which attracts insects to feed on it.
- The anthers are kept inside the flower so the insect brushes past them when it comes into the flower. Some flowers have very complex structures, which push the anthers onto the insect.
- The anthers make large, spiky or sticky pollen grains which stick to the body hairs of the insects that visit the flowers.

Insects are not the only animals that pollinate flowers. Hummingbirds and bats can also be attracted in similar ways.

Practical activity   Local insect pollinated flowers

Work in a group. Find as many different and interesting insect-pollinated flowers as you can in your local area and bring them into school. If you cannot collect any flowers, use the photographs on these pages and in topic 2.3 to help you.

- Draw each of the flowers.
- Look carefully for the way the flower is adapted to use insects to pollinate it.
- Label all of these features carefully on your drawings.

The pattern on this bee orchid looks like a female bee. When the male bee tries to mate with 'her' the anthers dip down and stick pollen to his head and body.

The amazing flower of the Indonesian Titan arum grows up to 10 m tall. It is also known as the corpse flower (or bunga bangkai – bunga means flower, while bangkai means corpse or body) because of the smell of rotting meat it produces to attract pollinating flies.

Key terms

- **nectar**

Summary questions

1   Make a large, colourful diagram of a typical insect pollinated flower and give labels and notes about the different features that you show.

2   Around the world, bees are dying of a new disease. People are worried that this will lead to a shortage of fruits such as apples and cherries. Why are they worried?

3   Using this book and any other sources you have available, list as many different animals as you can that act as pollinators for flowers, e.g. butterflies, bats.

Rice is the staple diet of almost half of the human population. The flowers are pollinated by the wind.

As you have seen, many plants rely on insects to pollinate them. This includes many of the plants that provide us with fruits to eat. However, many other plants around the world don't need insects at all. They rely on the wind to pollinate them. These plants are the grasses. They include all of the major cereals that make up our staple diet such as rice, maize and wheat.

Wind pollinated flowers

Insect pollinated flowers have adaptations which help them attract insects. Wind pollinated flowers do not need to attract the wind. However, they do have special features which increase the chance that the pollen will be blown away and will land on the stigma of another flower.

They look very different from the typical big, brightly coloured insect pollinated flower.

The features of wind pollinated flowers include:

- Small green flowers that can carry out photosynthesis so they get more energy.
- Anthers which hang outside the flower so the wind can easily blow the pollen away.
- Very tiny, smooth, light pollen which can be carried easily by the wind.
- Large amounts of pollen, as a lot of it will be wasted.
- A large, feathery and sometimes sticky stigma, which hangs outside the flower so it can pick up pollen from the air.

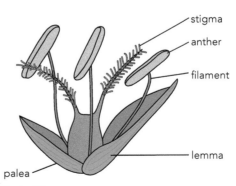

A typical wind pollinated flower

Practical activity   Local wind-pollinated flowers

Work in a group. Find as many different and interesting wind pollinated flowers as you can in your local area and bring them into school. If you cannot collect any flowers, use the photographs on these pages to help you.

- Draw each of the flowers. Look carefully for the way the flower is adapted to use the wind to pollinate it. Label all of these features on your drawings.

Maize flowers

Grass flowers

Self-pollination or cross-pollination?

Self-pollination is when pollen from a flower lands on the stigma of the same flower or onto the stigma of another flower on the same plant.

Cross-pollination is when pollen from one plant reaches the stigma of a flower on a completely different plant.

Each one has advantages and disadvantages.

	Advantages	Disadvantages
Self-pollination	Little risk that pollination will not take place because the anther and the stigma are in the same flower or on the same plant	No variety because all of the inherited information comes from the same plant
Cross-pollination	Variety – the inherited information from two different plants joins, which makes healthy offspring more likely	More risky – the pollen might not reach the stigma of another plant

Expert tips

Pollen grains are tiny; almost like dust. Seeds are larger. Some plants rely on the wind for transfer of pollen and also for seed dispersal. Do not confuse these two different stages of the life cycle.

Key terms

- **cross-pollination**
- **self-pollination**

Summary questions

1 Flower A produces small, smooth pollen grains. Flower B produces a small amount of big, spiky pollen grains. Which flower is insect pollinated and which is wind pollinated? What helped you to decide?

2 Work in a group. Make a model of a wind pollinated flower and an insect pollinated flower. Annotate the parts so people can understand the different ways in which they work.

3 Some flowers are more likely to be self-pollinated than others. Use secondary sources to help you find the answers to these questions:

   a) What are the advantages and disadvantages of self-pollination?

   b) What features do you think would make a flower more likely to be self-pollinated?

   c) How could a flower reduce the chances of being self-pollinated?

Pollination is very important for plants. As you have seen, flowers have many different ways of making sure that the pollen arrives safely on the stigma. But that is not the end of the story.

For the plant to reproduce successfully, the nucleus of the pollen grain must fuse with the nucleus of the ovule. To fuse, they must join together. The pollen grain is on the stigma, but the ovule is in the ovary at the bottom of the style. So how do they get together?

Pollen tubes

The top of the stigma makes a sticky sugar solution. The pollen grains stick onto it so they don't fall off or blow away once they have landed. The sugar syrup also makes **pollen tubes** grow out of the pollen grains. The tubes grow down the style and into the ovary. In the ovary the pollen tube grows into an ovule. Then the pollen nucleus moves down the tube. It fuses with the nucleus of the ovule. Now the ovule has been fertilised and a new plant can form.

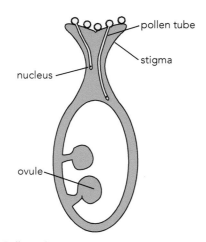

Pollen tubes growing into a carpel

Pollen tubes growing from pollen grains

Practical activity   Growing pollen tubes

Place a drop of sugar solution on a microscope slide.

Place a drop of water on another microscope slide.

Take an anther covered in pollen – or a flower with lots of pollen on its anthers – and shake it gently over both slides so some pollen falls into the drops of liquid.

Gently lower a coverslip onto both slides and leave them for about 20 minutes.

Use a microscope to look at the slides.

- Draw and label what you see. (You should see pollen tubes growing out of the pollen grains in the sugar solution.)

Once fertilisation has taken place, the fertilised ovule will develop into a seed. Once the ovules have been fertilised, most of the flower parts wither away and die, leaving the ovary behind. The ovary forms the fruit.

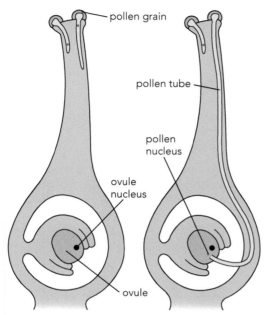

pollen grain

pollen tube

pollen nucleus

ovule nucleus

ovule

Fertilisation of an ovule

Practical activity   Does the sugar concentration affect the way pollen tubes grow?

You are given five different concentrations of sugar solution. You have to find out which concentration is best for growing pollen tubes.

- Plan how you would carry out this investigation.
- Describe how you would record your results and how you would display the results as clearly as possible.
- How will you decide which concentration of sugar is best for growing pollen tubes?

Show your plans to your teacher – you may be able to carry out the investigation in class.

Key terms

- **pollen tube**

Summary questions

1 What is the difference between pollination and fertilisation in a plant?

2 Draw an illustrated flow chart to show the sequence of events from pollination to the formation of a seed in the ovary.

3 The strength of the sugar solution might not be the only factor which affects fertilisation. Think of one other factor which might have an effect on how the pollen tubes grow and explain how you might investigate it.

After this topic you should be able to:

- describe a variety of fruits and seeds
- identify the different parts of a bean seed.

Once the ovules of a plant have been fertilised, they develop into seeds. The ovary forms a fruit which protects the seeds while they are developing and then helps in the process of **dispersal** (see topic 2.8). Some plants have only one seed in each fruit. Others have many. The seeds contain everything needed for a new plant to start growing.

Different fruits

Think of 10 different fruits – most of them are probably the sort of fruits we can eat. For example, the part of the cherry that we eat is the swollen ovary. The stone is the seed.

A tomato is also made up of the swollen ovary but it contains lots of seeds. In an orange the seeds are contained in the fleshy segments inside the leathery ovary wall. The pods of peas and beans are fruits, and so are groundnuts.

Different fruits have their seeds arranged and protected in different ways

Wind-pollinated plants such as rice and wheat also form fruits. The seeds are kept together until they are mature and ready to grow into new plants.

The grains on these ears of wheat fill up with stored food as they grow and mature

Practical activity   Investigating fruits

Collect as many different fruits together as you can. Cut them open and find the seed or seeds.

- Draw each fruit, label the seeds and then remove the seeds and count them.
- Record the number of seeds in each fruit and add the number to your drawing.

You will be using these drawings in topic 2.8 as well.

The structure of a seed

Each seed contains everything needed to grow into a new plant. Inside the seed there is a tiny new plant called the **embryo plant**. The seed has a tough coat around it to protect the embryo plant. It also contains a store of food. This supports the tiny new plant until it grows out of the seed and can make food for itself by photosynthesis. This store of food is usually rich in starch and lipids which contain lots of energy for the new plant.

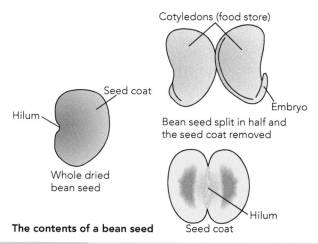

The contents of a bean seed

Practical activity   Investigating seeds

Collect a bean seed which has been soaked overnight in water.

- Look at the outside of your bean. Draw and label it. See if you can see the shape of the root of the embryo plant under the tough skin. Can you find the scar where the seed was attached to the fruit and the tiny hole where the water gets in to reach the embryo plant?
- Carefully remove the outer coat and gently break the bean into two parts. Look for the embryo plant. You may be able to identify the embryo root and shoot.
- Take one part of the cotyledons, which are the food store, and crush it slightly. Add iodine solution and see if you can identify starch.

Germinating seeds

A seed needs water, warmth and oxygen to grow. Germination takes place as the embryo plant starts to use up the food stores and grow. The new root is the first part to break through the tough outer coat, followed by the shoot. The roots go down into the soil and quickly begin to take up water. As the shoot grows above the soil, the tiny leaves open and the plant starts to photosynthesise.

Key terms

- **dispersal**
- **embryo plant**

Practical activity   Investigating how seeds grow

The seed may fall into the ground any way up.

- Grow some soaked bean seeds in a jar with a roll of blotting paper to hold the beans in place.
- Pour water into the bottom of the jar.
- Put the beans in the jar different ways up and observe what happens as they grow.

Summary questions

1 Make a poster showing as many different types of fruits as you can. For each fruit show where the seeds are found.

2 Why is it important that the seeds form inside a fruit?

3 Draw and label a diagram of a bean seed. Then draw a series of diagrams to show what happens as the seed germinates.

Learning outcomes

After this topic you should be able to:

- explain why seeds need to be dispersed as far as possible from the parent plant
- make observations and measurements
- interpret results using your scientific knowledge
- draw conclusions and evaluate the methods used.

Plants need lots of things to grow. They need plenty of light. They need water and mineral salts from the soil. This means there are possible problems for plants when they reproduce and make seeds which will grow into small new plants.

Competition in plants

If plants are crowded, they compete with each other for space, for light, for water and for minerals from the soil. When seeds land on the soil and start to grow, the plants that grow fastest will do best. If all of the seeds of a plant land together on the soil, they will compete against each other and some of them will not grow very well and may die.

If the seeds of a plant land close to the parent plant, the parent plant will be in direct competition with its own seedlings. The adult plant is large and strong, so it will take most of the water, light and minerals. It will deprive its own offspring of the things they need to grow. To stop this happening, most plants have ways of dispersing their seeds, spreading them as far away as possible from the parent plant (see topic 2.9).

Investigating overcrowding

Scientists have carried out many investigations into the effect of overcrowding on the growth of plants. This is done by growing two sets of plants, one planted very close together and the other planted spread apart. It is important to keep all the other conditions – light intensity, temperature, amount of water and minerals – the same for both sets of plants. This means any differences in the way the plants grow are the result of overcrowding and competition for resources.

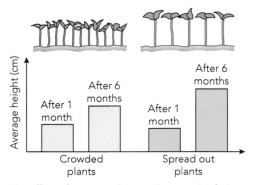

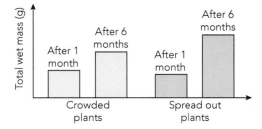

The effect of overcrowding on the growth of plants

Practical activity   Investigating the effect of overcrowding on growth

You are going to investigate the effect of overcrowding on the growth of plants using cress or mung seedlings.

Set up two dishes or lids with cotton wool soaked in water.

On one dish put a thick layer of cress or mung seeds so you cannot see the cotton wool.

On the other dish scatter just a few seeds with plenty of space between them.

Leave the seeds for a week to grow. Make sure they are given the same amount of water every day, and that they are in the same place in the classroom so they get the same temperature and the same amount of light.

After a week, take five cress plants from each dish. Examine them carefully. Measure the length of their stems, the size of their leaves (you can use graph paper to let you measure the area of the leaves or just measure the length and breadth) and the amount of root they have formed.

- Record your results in a table.

- Work out the mean measurements for each group of plants and plot bar charts to compare them.

- How has overcrowding affected the growth of these plants?

- What are the limitations of your experiment? How could you improve it?

When farmers plant a crop, they want to grow as many plants as possible. But if they put the seeds too close together they will compete and none of them will grow well.

Summary questions

1 How does overcrowding affect the growth of plants?

2 In the experiment shown in the figure opposite explain why the mean wet mass and the mean height were measured.

3 a) In the experimental results shown in the figure opposite the plants grown in overcrowded conditions were taller after one month than the plants grown in spread out conditions. Suggest an explanation for this.

   b) After six months, the spaced out plants were taller than the plants grown in crowded conditions. Suggest an explanation for this.

Dispersal mechanisms

A fruit is something that protects the seeds of a plant as they form and then helps in the dispersal of the seeds. Seeds need to be dispersed (spread as far as possible from their parent plant and the other seeds). That way they avoid competition with their parent and with all the other seedlings that will grow.

How are seeds dispersed?

There are four main ways in which seeds are dispersed. It all depends on the type of fruit the seeds are hidden in! They include:

- animals
- wind
- water
- self dispersal.

Animals

Some fruits, such as cherries, grapes, oranges and apples, are sweet and fleshy. These fruits are eaten by birds and other animals (including people). The seeds have a very hard coat so they are not digested. Some time after they are eaten, the seeds pass out of the animal in the faeces. The seeds land on the soil a long way from the parent plant, with their own supply of manure.

This bird is getting a meal – but later it will disperse the seeds of this plant

Some fruits are sticky or have hooks on them. These fruits get stuck to the fur of an animal. Later, when the animal grooms itself, the fruit falls off onto the ground and the seeds can grow.

Sometimes the ovary wall becomes very hard, forming a nut. Animals carry the nuts away and hide them to eat later. The animals often forget where they have hidden the nuts, so later the seeds can begin to grow.

Wind

Some seeds are spread by the wind. They may have wings which make them spin like helicopters as they fall from the tree. This means they are carried further by the wind. Other seeds are very small, with fluffy parachutes which help them to be carried long distances by the wind.

Winged seeds are dispersed by the wind

1 You may collect or be given some winged fruits from a local tree – you are going to investigate how effective the wings are at helping the dispersal of the seeds by dropping the fruits from a height and seeing how far they travel.

Work in groups of three – one to drop the fruit, one to measure the distance it flies and one to record the results.

Drop the same fruit three times from the same height and measure the distance it travels. Find the mean distance flown by the fruit. Remove the wings and repeat the experiment.

- What does this show you about the importance of wings in the dispersal of seeds?

2 Make a paper model of a fruit with a seed. It must have two or more wings and a weighted centre to mimic the seed.

Change different factors about your fruit and decide what combination flies best.

- Make a presentation about your seed design and write a conclusion describing the features that enable fruits to travel furthest.

Water

Coconuts are one of the biggest seeds. They grow on palms near the sea. The coconuts are covered in fibres which trap the air. This means they can float long distances on the sea before they are washed up onto another shore and can start to grow.

Coconuts are the best-known seed dispersed by water

Expert tips

This section is about seed dispersal, not about pollination. Pollination occurs before fertilisation; seed dispersal occurs after fertilisation.

Self dispersal

Some fruits explode as they dry out. This sends the seeds flying though the air away from the parent plant. Peas and beans are examples of explosive seeds, although we usually harvest them and eat them before they get a chance to explode.

Summary questions

1 Why is it a good thing for seeds to grow as far away as possible from their parent plant?

2 Using the information on these pages and any secondary sources you have, make a big poster to show the different ways that seeds can be dispersed, showing lots of different examples.

3 Explain EITHER how you would investigate the best colour for berries to encourage birds to eat them OR how wind affects the distance that parachute seeds are dispersed.

1 The diagram shows a vertical section through an insect pollinated flower.

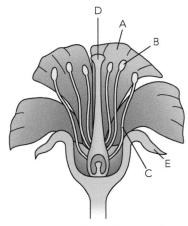

a Name the parts of the flower labelled **A** to **E**. [5]

b The diagram shows the ovary and ovules of this flower. Use the diagram above to state the difference between the ovary and the ovules. You may make a labelled drawing if you wish. [2]

c Describe three ways in which insect pollinated flowers attract insects. [3]

d What are the advantages of transferring pollen from one plant to another of the same species? [3]

2 The following events occur during sexual reproduction in flowering plants.

A pollination; **B** growth of embryos; **C** dispersal of seeds; **D** growth of pollen tubes; **E** fertilisation; **F** flower formation; **G** production of pollen; **H** germination of seeds.

a The first event is **F** and the last one is **H**. Write down the sequence in which the rest of the events occur. Use the letters above starting at **F** and ending with **H**. [2]

b Explain the difference between each of the following pairs of terms:

  i   wind pollination and insect pollination

  ii   cross-pollination and self-pollination

  iii   ovule and pollen grain

  iv   gamete and embryo

  v   pollination and seed dispersal

  vi   pollination and fertilisation [12]

c This photograph shows some flowers of a species of grass.

Explain how the features of the flower that you can see in the photograph help the transfer of pollen during pollination. [3]

3 Some Polish scientists studied how flowers of two species of deadnettle were pollinated. They counted the number of visits made by different insects to flowers of the red deadnettle and the white deadnettle growing in the same area. These pie charts show their results.

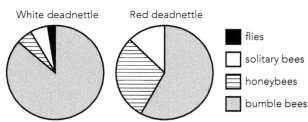

a i   Which type of insect paid most visits to the deadnettle flowers? [1]

  ii   Which type of insect only visited one of the species? [1]

  iii   The scientists saw many of the insects visiting both species over short periods of time. Suggest why this did not lead to the production of cross breeds between the two species. [2]

Honeybees have been observed making several trips from their hive to the same group of flowers. Other insects do not live in hives and visit the flowers of many different plants over short periods of time.

b Suggest how the scientists could find out how many different types of flower each species of insect visits over the period of an hour. *[3]*

4 Samira was studying fruits and seeds. She made drawings of sections cut through a tomato and a broad bean.

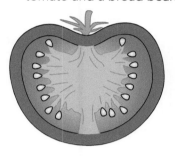
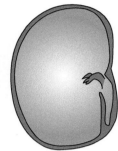

a Copy the drawing of the tomato and label the seeds, the fruit wall and the sepals. *[4]*

b Copy the drawing of the broad bean and label the embryo, cotyledon and testa. *[4]*

c Broad bean seeds contain starch, but tomatoes do not. Describe how you would confirm that this is true. *[3]*

d Explain how you would find out if tomatoes and broad bean seeds contain sugars. *[6]*

5 A class of students collected many different fruits from the grounds around their school. Some of these fruits are shown here.

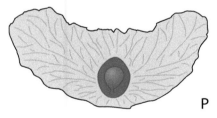

P

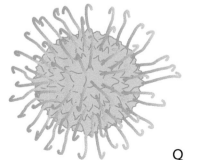

Q

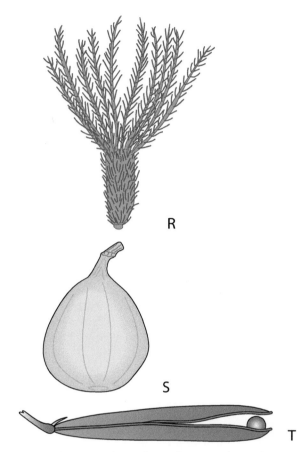

R

S

T

a Suggest how these fruits and seeds labelled **P** to **T** are dispersed. *[4]*

b Explain why fruit and seed dispersal is important for plants. *[3]*

c The seeds of most plants do not germinate immediately after dispersal. They can remain dormant for many years. Explain how dormancy of seeds is an advantage for plants. *[3]*

The seeds of some plants, such as grasses, can be dried and kept in cold storage in seed banks for many years. Important tropical crops, such as cocoa and coconut, have seeds that cannot be kept in storage. Many tropical crops are at risk from pests and diseases.

d Explain why it is important to keep seeds in seed banks. *[3]*

e Suggest how scientists can make sure tropical crops such as cocoa can be kept for the future even though their seeds cannot be stored. *[2]*

Science *in context!*

Big cats and the Masai Mara

The Masai Mara is a large and famous game reserve in Kenya. It is an area of trees, scrub and savannah, which is home to many animals and also to the Masai people. The Masai have lived in the area for many years, rearing cattle. They have a detailed knowledge of the country and the wildlife. One of the best known of the Masai is Jackson Looseyia. He was trained by his father who was head ranger to understand the ways of the animals and how to find them. He now leads safaris and also works on television programmes to bring the wonders of the Masai Mara to people around the world.

Some of the most dramatic animals in the Mara are the cats. They are able to live in the same area without competing with each other because they have different ways of life:

Jackson Looseyia

Lions

Cheetah

Leopard

Serval

Lions, cheetahs, leopards and serval – just some of the cats found in the Masai Mara

Lions are big, social animals living in groups called prides. The females hunt together which means they can bring down large prey such as zebra and wildebeest. They hunt during the day and the whole pride feeds on the kill.

Cheetahs are smaller and lighter than lions. They hunt during the day and rely on a very fast burst of speed to catch their prey – they are the fastest land animals on the Earth. Their prey is usually smaller and slower than they are, such as Thomson's gazelle, and they eat it immediately before it is stolen by bigger animals.

Leopards live and hunt alone and often at night in the Mara. They climb trees and often store their kills in a tree away from other predators. They eat a wide range of prey, which includes beetles, monkeys and apes and giant elands. They stalk silently and pounce on their prey at the last minute without a chase. If there are lions hunting big prey in the same area, leopards will usually take smaller prey than if there are no lions in the area.

Servals are smaller than lions, leopards and cheetahs. They hunt mainly at night. Their prey is rodents – rats and mice – birds, fish and reptiles.

In this chapter you will find out about the interrelationships between living organisms. You will also study genetics and the importance of natural selection in evolution.

Key points

- Classification keys are useful for identifying and classifying organisms.
- Food chains are simple models of the feeding relationships between different organisms. They always begin with a plant or other organism that can photosynthesise.
- Food webs are made up of many different food chains linked together. They give a more accurate picture of the feeding relationships in a habitat.
- Not all of the biomass in an organism is passed on to the next organism in the food chain.
- Energy flows through a food chain from the Sun eventually to the decomposers. Energy is lost to the environment at each stage of the chain.
- The decomposers are bacteria and fungi. They break down the waste from animals and the bodies of dead animals and plants, returning minerals to the soil.
- A food chain or web consists of different trophic (energy) levels – producers, primary consumers, secondary consumers, tertiary consumers and the decomposers.
- The genetic material is DNA in the form of chromosomes found in the nucleus of the cell.
- Genes carry the information about individual characteristics. Each chromosome contains many genes.
- The sex cells contain half the number of chromosomes. When they combine, the new cell has a combination of genes from both parents. This introduces variation.
- In selective breeding you choose parent organisms with a characteristic you want and breed them to get offspring which have the desired characteristic.
- Selective breeding leads to the formation of new varieties. It has been used to improve crop plants and domestic animals for centuries.
- In nature, animals and plants with characteristics that help them to survive and therefore breed in their particular environment will pass on their genes successfully. This is natural selection.
- Natural selection drives the development of different species which live in the same environment but do not compete with each other for resources. New species also form when organisms are isolated geographically and cannot breed.
- The theory of natural selection was developed by Charles Darwin.
- He collected evidence for many years before presenting his theory to other scientists.
- The way scientists work has changed and developed over time. Evidence and clear recording of results has become more and more important in the development of scientific theories.

We classify living organisms by looking at the things that are similar between them, and also the things that are different. When we find a strange organism, we need to be able to tell which group it fits into. This is where classification **keys** are very useful.

Practical activity   Classifying organisms

You learnt about classification in Stage 7 – how much can you remember?

Work in a small group and write your answers to the tasks below on a sheet of paper.

- List the main groups of vertebrate animals.
- List the main groups of invertebrate animals.
- List the main groups of plants.

Keys

We can find out what an organism is by using a key. A key is a tool you can use to help you identify and classify organisms. Opposite is a key you could use on a visit to an aquarium to identify animals from the oceans.

Practical activity   Identifying fish

Choose the fish you want to identify from pictures A–F. Then in the key at the top of page 49 until you identify your fish answer yes or no to the questions.

A

B

C

D

E

F

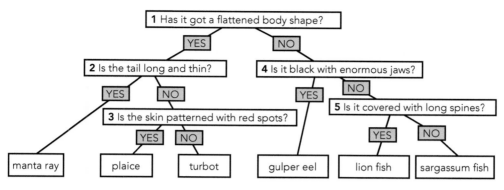

Biologists often use keys when they are working out in the field, observing plants and animals in their natural surroundings. To take up less space, keys are usually printed as a series of questions and answers. Here is the branching key you have just used to identify the fish shown as a table of questions. See if you can still use it.

1	Has it got a flattened body shape?	**Yes:** go to question **2**	**No:** go to question **4**
2	Is the tail long and thin?	**Yes:** <u>manta ray</u>	**No:** go to question **3**
3	Is the skin patterned with red spots?	**Yes:** <u>plaice</u>	**No:** <u>turbot</u>
4	Is it black with enormous jaws?	**Yes:** <u>gulper eel</u>	**No:** go to question **5**
5	Is it covered with long spines?	**Yes:** <u>lionfish</u>	**No:** <u>sargassum fish</u>

Practical activity   Making a key

Either: Go out into the area surrounding your school and collect five biological specimens, e.g. leaves/flowers/fruits from different plants, different invertebrates.

Or: Look at the five specimens given to you by your teacher.

- Draw your specimens.
- Make a branching key to identify your specimens.

Try out your key on a friend.

- If your key works, turn it into a series of questions in a table. If it doesn't work, change your questions until the key can be used successfully and then make it into a table.

Key terms

- **key**

Summary questions

1. a) What is a classification key?
   b) Why are keys so useful to biologists?

2. The vertebrates can be divided into five groups: fish, amphibians, reptiles, birds and mammals.
   a) Make a branching key that divides the vertebrates up into these five groups. Think about features such as the number of legs, the presence of gills or wings and the sort of skin the animal has.
   b) Write out your key in a table, as shown above.

The plants and animals that live in similar habitats are linked by food chains. Remember that food chains always start with a plant. That is because they are the **producers**, making food by photosynthesis.

A simple food chain seen in many countries around the world: grass → rabbit → fox

Food webs

A food chain is a very simple model of what happens in the natural world. Think of the food you ate yesterday. You had more than one type of food. In the same way animals eat lots of different things.

Herbivores eat plants, and most of them eat lots of different types of plants in a day. **Carnivores**, such as the fox in our food chain above, eat more than one type of animal prey. Foxes will eat rabbits, eggs, small birds, lizards, beetles, and more. To try and give a more accurate picture of the feeding relationships in a habitat, we link lots of food chains together to make a **food web**. Sometimes it takes a lot of careful observation to decide exactly what an animal actually eats!

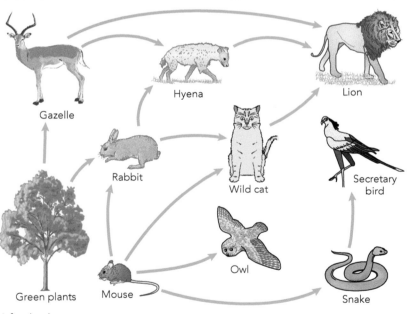

A food web

Practical activity   Finding food chains in food webs

Look at the food web shown in the figure above.

- Find as many food chains as possible in the web and write them out in a list.

Use some of the organisms in your local area to build up the biggest food web you can.

Pyramids of numbers and biomass

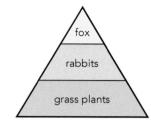

Thousands of grass plants feed only a few rabbits and these rabbits would feed only one fox. Knowing the numbers of organisms in a food chain, you can organise them in a **pyramid of numbers**.

However, when an animal eats a plant, not all of the food in the plant ends up as living material in the animal. Some of the plant material cannot be digested. Some of it is also used up for other things such as energy for moving about.

When an animal is eaten itself, it passes along only the food that has become part of its body. The next organism gets only the small amount of material that was turned into new biological material. This biological material is called **biomass**. This means that as you go along a food chain, there is less and less biomass, however many organisms are involved. This can also be organised into a pyramid of biomass.

A pyramid of number

Energy flow through a food web

In the same way, food chains show the flow of energy through organisms from the Sun to the decomposers. Just as the biomass passed on at each level gets less and less, so the amount of energy that flows through a food chain gets less at each level.

Much of the energy taken in by an animal is used to keep warm or for moving around. Some of the energy in food is passed out of the animals in urine and faeces. Only about 10% of the energy taken in is actually stored in the body of an animal and so this is all that can flow on through the chain.

Key terms

- **carnivore**
- **food web**
- **herbivore**
- **producer**
- **pyramid of numbers**
- **biomass**

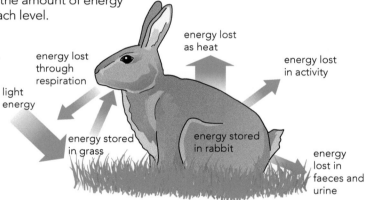

Only the energy stored in the rabbit will be available to flow on to a fox if the rabbit is eaten

---

Summary questions

1 Why is the bottom level of a food web or a pyramid of numbers always filled with plants?

2 Winkles and limpets eat seaweed. Winkles are eaten by octopuses and crabs. Crabs and starfish eat limpets. Seals eat crabs and octopuses. Octopuses eat crabs. Seagulls eat crabs and starfish. Killer whales eat seals.

   a) Use this information to draw a food web.

   b) Name the **i)** producer, **ii)** the herbivores, **iii)** the carnivores, in this food web.

   c) If a disease killed off most of the crabs, what effect do you think this would have on the rest of the food web?

There is one important group of living organisms which are often not shown in food chains and food webs. These are the **decomposers**.

The decomposers and recycling

Animals and plants in food chains and webs die. While they are alive, animals also make waste products such as faeces and urine. If all of this waste and dead bodies built up in the environment, there would not be room for any new organisms.

This overcrowding doesn't happen because the bodies and waste are broken down by decomposers. These include bacteria and fungi, as well as animals such as woodlice and earthworms. These decomposers feed on dead bodies and waste and use it as food. They break it down during respiration to get the energy they need.

The decomposers break down organisms until there is little or nothing left

Practical activity · Observing decomposers at work

Place a fruit such as an orange or a tomato in a sealed container, e.g. a jar with a lid. You can use a piece of damp bread instead of a fruit.

Leave the jar in a warm place and watch what happens over time.

- Make a record of what you see with written observations, drawings or photographs.

Sometimes things decompose very quickly and sometimes it is very slow.

- Plan how you could change this basic experiment to investigate the conditions that affect how fast decomposers work.

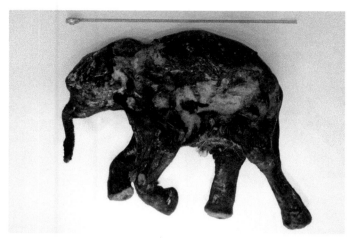

Sometimes decomposition is very slow. The body of this mammoth was discovered in frozen ground thousands of years after it had died.

The bodies of animals and plants contain lots of minerals. When the decomposers break them down, these minerals are released back into the soil. Then they can be taken up by the roots of plants to be used to make more plant cells and to pass through the food chains and food webs again.

Trophic levels

As you can see, a food chain should always end with the decomposers. The energy that flows through the food chain ends up in the bodies of the decomposers or is released as heat energy. You can look at each stage of a chain or web as an energy level. The amount of energy contained in the organisms is less at every stage:

- the plants are producers
- the herbivores are **primary consumers**
- the carnivores that eat the herbivores are **secondary consumers**
- the carnivores that eat other carnivores are **tertiary consumers**
- the microorganisms and other animals that eat all of the different types of organisms are the decomposers.

You can label any food chain or web with these **trophic levels**.

 IGCSE Link...
You will learn about the roles of decomposers in recycling the elements carbon and nitrogen so that plants and animals do not run out of these important elements.

Key terms

- **decomposer**
- **primary consumer**
- **secondary consumer**
- **tertiary consumer**
- **trophic level**

Summary questions

1 Draw a diagram to show how the decomposers recycle minerals from the bodies of organisms and make them available again to plants.

2 The baby mammoth shown in the photograph above was almost complete. Suggest reasons why it had not been broken down by decomposers.

3 Revisit your local food web and re-label all the organisms by their trophic level. You could use colour coding.

Genes and genetics

Galileo

Kind

Frankel

Noble Mission (right)

The racehorses Galileo and Kind are the parents of both Frankel and Noble Mission

Look closely at these horses. The offspring and their parents have some similarities – look at their colours, their body shapes and their markings. Now look again at Frankel and Noble Mission. Although they were born to the same mother and father, they look quite different. And while Frankel is one of the most famous racehorses ever, his brother is not quite as good.

Information in the genes

You know that every cell contains a nucleus. The nucleus contains the genetic material in the form of a set of **genes**. Each gene has the information needed to control the development of a feature such as dimples or eye colour.

The genes are carried on pairs of **chromosomes** in the cell nucleus. The chromosomes are made of **DNA** and each contains thousands of genes. Humans have 46 chromosomes arranged in 23 pairs; horses have 64 chromosomes arranged in 32 pairs.

When the sex cells or gametes are made, they only contain half of the chromosomes. When a male and female gamete join together, the two half sets of chromosomes make

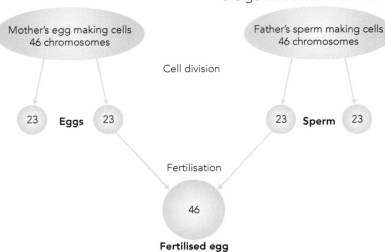

Mother's egg making cells 46 chromosomes

Father's sperm making cells 46 chromosomes

Cell division

23 **Eggs** 23

23 **Sperm** 23

Fertilisation

46

Fertilised egg

The way gametes are formed and join to form a new, different individual a full set again. The offspring gets one copy of each gene from both parents. As the fertilised egg or ovule divides to form an embryo, exact copies of the new set of chromosomes are made again and again.

Variety, which is passed down from parents to offspring, can be seen everywhere. Plants pass on their genetic plans in the seeds that result when their male and female gametes fuse. Animals pass on information in the same way as people, when egg and sperm join together.

Brothers and sisters are different from each other because the mixture of genetic information is different in every egg and sperm. This means every new person is different. The only exception is in the case of identical twins. Here the egg and sperm join, begin to develop into a baby and then split to form two people whose cells have the same information.

PARENT'S GENES

The parent has ten genes arranged in two sets.

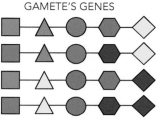
GAMETE'S GENES

Here are four different combinations of five genes in the gamete. How many more are there?

Identical twins have identical genetic material

The genes of the parent organism can be combined in many different ways to produce variety in the offspring

---

Practical activity   Inherited features

Think about your head:

- How many different visible features on your head can you list that you think are inherited?
- What things about the appearance of your head are not inherited?

Summary questions

1 Give a definition of the following three terms:
   a) gene   b) chromosome   c) DNA.

2 a) Explain why the children in a family often look similar.
   b) Explain why the children in a family usually have differences between them.
   c) Explain why sometimes children in a family may look exactly the same.

3 Make a poster of your own family or of a family you know using photos to show the similarities and differences between the family members.

You have seen that variation happens as a result of inheritance. But not all variation happens as a result of chance. People often become involved to obtain the plant or animal of their choice.

Selective breeding

Fancy goldfish like these are the of result selective breeding

Around the world people enjoy watching goldfish swim in tanks. However goldfish are not found in all these amazing forms and colours by accident. People have used **selective breeding** to obtain the features they want in these different types of fish.

In selective breeding you choose as parents two animals or plants which have the feature you want to improve and breed them together. You select individuals that show this feature and breed them together. You select the offspring that show an improvement in the feature and breed from them again. Continue the process until you have the organism you want.

Food for all

Thousands of years ago, our ancestors collected the seeds of grass plants to eat. Gradually, they started sowing their own grass plants. They chose the biggest, fattest seeds to sow. Although they didn't know it, they were selecting the alleles most useful to people. Over centuries, the early grass plants became all the cereal crops we use today, which have lots of large seeds for us to eat.

Selective breeding is very powerful. Look at the brassicas in the figure on page 57. All of the members of the cabbage family come originally from wild mustard plants. However over hundreds of years people selected parent plants with the features they wanted to breed from and now we have some very different vegetables.

Remember that in the past people did not understand about genes and genetics. They could only work by trial and error. Now we understand how characteristics are passed from parents to their offspring, selective breeding gets results more effectively and faster.

By selective breeding, people have developed crops which can cope well with extreme conditions e.g. little water, but still produce a lot of grain. They have developed cereal crops with much shorter stems, which are much less likely to be damaged by storms. These developments in selective breeding can help to produce more

By looking at the original plants (on the left and centre) and a modern maize cob (on the right), you can see how selective breeding has increased the size of the seed head. This gives us much more food per plant.

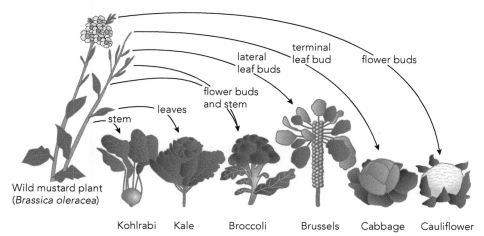

stem — leaves — flower buds and stem — lateral leaf buds — terminal leaf bud — flower buds

Wild mustard plant (*Brassica oleracea*)

Kohlrabi   Kale   Broccoli   Brussels   Cabbage   Cauliflower

Selective breeding has produced many different brassicas by selecting for different parts of the plant

food in developing countries. Producing cereals which give bigger yields on the same land is also an important result of selective breeding over the years.

However, for selective breeding to work it is important to breed only the best animals or plants with the characteristics you want. It is important to get lots of farmers working together so that the animals or plants do not get too inbred.

Practical activity   Select the best

Think about a fruit or vegetable you like. What would make it better? Would you prefer a smoother skin, or a sweeter taste – or would it look more delicious if it was a different colour?

Or what about your favourite flower – would you like to see it in a different colour?

If you have animals at home as pets or to provide food – what would make them better? Could they be bigger or smaller? Could they give more milk or lose less hair on the furniture?

- Decide how you would like to improve a plant or animal of your choice. Then make a poster showing how you would do it, and what your new organism would look like.

Summary questions

1 What is selective breeding?

2 How does selective breeding help us to improve our crops and farm animals?

3 Here are some examples of organisms produced by selective breeding:
   i) sheep with extra thick and woolly fleece
   ii) potato plants that produce large smooth potatoes
   iii) wheat plants that are resistant to the plant disease rust
   iv) cows with udders that can produce large volumes of milk.
      a) Why do you think each of these organisms has been produced?
      b) How do these examples show the importance of selective breeding for people?
      c) Why is it easier to carry out selective breeding now than it was a hundred years ago?

3.6 Natural selection in nature

IGCSE Link...

You can see that these antelopes look different from each other. At IGCSE you will learn how these differences come about.

When we use selective breeding to improve our crops or domestic animals, we are borrowing a process that has worked for millions of years in nature. All of the different wild animals and plants in the world exist as a result of selective breeding. But in nature, selection takes place by survival. It is known as **natural selection**.

What is natural selection?

Any group of animals or plants will have lots of characteristics in common, but there will always be some differences. When life is good and there is plenty to eat, these differences do not matter. However things can get difficult. There might not be enough food, a new disease may attack the organisms, or new predators may arrive. They have to compete with each other to survive in these more difficult conditions.

Some of the animals or plants will have **adaptations**, which mean they can survive better than others. There are always more organisms than an area can support, so the organisms with the best adaptations are most likely to reproduce and pass on their genes. Gradually, the population will change until all of them have the characteristic which allowed some of them to survive. This natural selection is also known as '**the survival of the fittest**'.

Natural selection in action

There are 72 different species of antelope. If they all ate the same type of grass, there would be competition between them for food and many would die out. But if natural selection results in antelope feeding on different types of plants, they will have more food and less competition. They will be able to rear their offspring successfully. Eventually there will be many herds of antelope eating quite different plants.

Grant's gazelles live in dry conditions and mainly browse on shrubs and plants from dry areas. Thomson's gazelles rely more on water and mainly eat grass. They are very similar but do not compete for the same food.

Sitatunga have long, splayed out feet, which let them walk on marshy ground to feed on swamp plants, so they do not compete with other antelope.

The greater kudu sleeps hidden in scrub all day and comes out to feed at night. Its behaviour is adapted to avoid disturbance by people and so it can eat with little competition.

There are many more ways in which natural selection has changed antelope so that they are perfectly adapted to a particular way of life. In each case the adaptations are passed down the generations.

The tiny dik-dik is only 30–40 cm high so it can escape predators and hide in the bushes. In contrast, the giant eland is 130–180 cm high at the shoulder – so large that predators think twice before attacking it.

Key terms

- **adaptation**
- **natural selection**
- **the survival of the fittest**

Summary questions

1 **a)** What is an adaptation?

   b) What is natural selection and how does it result in different types of animals and plants?

2 List some of the things that can drive natural selection to take place.

3 Use secondary sources to discover some of the animals and plants that result from natural selection on an island such as Australia or Madagascar. Make a presentation to show others about these animals and how natural selection has affected them.

Our modern ideas about how all the different forms of living organisms have developed started with the work of one of the most famous scientists of all time – **Charles Darwin**. Darwin set out in 1831 as the ship's naturalist on HMS *Beagle*. He was only 22 years old when he set off on the voyage to South America and the South Sea Islands.

Darwin and the voyage of the Beagle

Darwin planned to study geology on the trip. But as the voyage went on he became as excited by his collection of animals and plants as by his rock samples.

In South America, Darwin discovered a new form of the common rhea, an ostrich-like flightless bird. Two different types of the same bird living in slightly different areas made Darwin think. The Darwin's rhea feeds on shrubs and even desert plants as well as grass, while the greater rhea feeds only on grassland. Darwin only noticed the new species when he was half way though eating it! He sent the remains back to England to be studied.

The greater rhea can be up to 1.80 m tall, but the lesser or Darwin's rhea is only 0.9 m tall. These birds have evolved by natural selection and feed in different areas.

On the Galapagos Islands in South America, Darwin was amazed by the variety of animals. He noticed that they differed from island to island. Darwin found strong similarities between types of finches, iguanas and tortoises on the different islands. Yet each was different and adapted to make the most of local conditions.

Darwin collected huge numbers of specimens of animals and plants during the voyage. He also made detailed drawings and kept written observations. The long journey home gave him plenty of time to think about what he had seen. Charles Darwin returned home after five years with some new and different ideas forming in his mind.

After he got home to England, Darwin spent the next 20 years working on his ideas. He worked hard to put together evidence that supported his ideas. This was unusual in those days. Darwin looked at the different animals and plants he had found on his travels and how they were related to each other. He also did lots of work breeding pigeons and showing how artificial selection worked. In spite of all this work it took many years before some people accepted his ideas.

Darwin was very impressed by the giant tortoises he found on the Galapagos Islands. The tortoises on each island had different shaped shells and a slightly different way of life. Darwin made detailed drawings of them all.

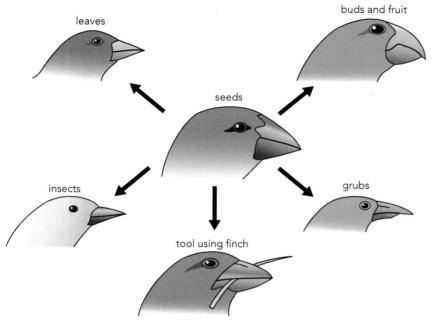

leaves

buds and fruit

seeds

insects

grubs

tool using finch

Darwin's finches show different types of beaks, which allow them to eat different types of food so they don't compete with each other

Darwin's theory of evolution by **natural selection** is that all living organisms have evolved from simpler life forms. This evolution has come about by a process of natural selection. Reproduction always gives more offspring than the environment can support. Only those that have inherited features most suited to their environment – the 'fittest' – will survive. When they breed, they pass on the genes for those useful inherited characteristics to their offspring. This is natural selection.

Key terms

- **Charles Darwin**
- **natural selection**

Practical activity   Finding out about Lamarck

Shortly before Darwin, a scientist called Jean-Baptiste Lamarck had an idea about how different species came about. It was called the theory of acquired characteristics. However Lamarck did not have any evidence to support his idea and people could see it did not work.

Use books and other secondary sources such as the internet to find out about the work of Lamarck.

- Make a poster to explain the ideas of Lamarck and why they are not correct.

Summary questions

1 Explain what is meant by **a)** natural selection and **b)** evolution.

2 Suggest reasons why it may have taken some people a long time to accept Darwin's ideas and some do not accept them (you may want to discuss this question in a group).

3 Make a big illustrated timeline showing some of the main events in Darwin's life (use secondary sources to help you).

3.8   Natural selection in action

Learning outcomes

After this topic you should be able to:

- describe examples of natural selection in action
- explain the importance of evidence and explanations in science.

Expert tips

Scientists have found other ways in which pollution has changed the environment for organisms so that natural selection has prompted changes like those that occurred to the peppered moth.

When Darwin suggested how evolution takes place, no-one knew about genes. Darwin simply observed that useful inherited characteristics were passed on. Today we know it is useful genes which are passed from parents to their offspring in natural selection. Here are some examples of natural selection in action.

The story of the peppered moth

The **peppered moth** lives in temperate countries such as the UK. It flies at night and hides during the day on the bark of trees. In the early 19th century most peppered moths were white speckled with black. They were almost invisible against the lichen-covered bark of many trees. This made it hard for birds to find and eat them. Any dark moths that appeared were easily seen and eaten so they did not often live to reproduce.

After the Industrial Revolution many trees became darkened with soot from the factory smoke. By the end of the 19th century up to 95% of the peppered moths in some areas were black. Now the pale peppered moths were easy for the birds to see and eat. As a result of natural selection, the darker moths took over.

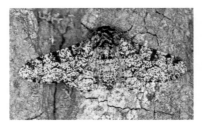 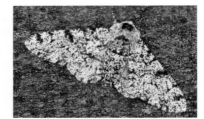

Look carefully at these pictures and spot the moths on both of them. If you were a hungry bird, which moth would you see most easily?

In the middle of the 20th century the air was cleaned up and then the trees became clean again too. The black moths stood out and were eaten in large numbers. The pale moths were well camouflaged and survived to reproduce. Today most peppered moths in the UK are pale again, with just a few black moths appearing, if the birds don't eat them first.

Bernard Kettlewell did the original research and recorded the colour of moth populations all over the UK. Recently, some modern scientists questioned his research but in 2007, new evidence was published by Professor Majerus at Cambridge University, UK, which showed clearly that Kettlewell's ideas were correct. Professor Majerus filmed many birds finding and eating the dark moths on light trees – showing everyone an example of natural selection in action.

Antibiotic resistance

Natural selection is not something that just happened in the past. It is still happening today. For example, some infections are caused by bacteria; e.g. tonsillitis and tuberculosis. Antibiotics are medicines that kill bacteria and make you better. But sometimes they don't work because they are **antibiotic resistant**. This is another example of natural selection taking place:

- A small number of the bacteria making you ill will be resistant to the antibiotic you are given.
- The antibiotic will kill all the normal bacteria. If you keep taking the antibiotic to the end of the course it will probably kill ALL of the bacteria.
- If you stop taking the medicine too soon, some of the resistant bacteria will survive. They will reproduce and form a big colony of antibiotic resistant bacteria. This is natural selection.
- If they infect someone else, or grow again in you, the antibiotics will not kill them and cure the disease.
- A different antibiotic will probably kill them, but as a result of natural selection some bacteria have become resistant to almost all the antibiotics we have.

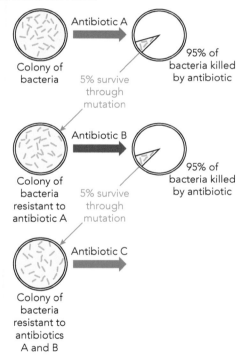

Natural selection and the development of antibiotic resistance. The process continues with the introduction of new antibiotics

Key terms

- **antibiotic resistance**
- **peppered moth**

Summary questions

1 Explain what is meant by natural selection using the peppered moth as an example.

2 Make a flow chart to show how bacteria develop resistance to antibiotics.

3 Using secondary sources, find another example of natural selection and write a case study of what has happened to present to the class.

Moorea is a tiny beautiful island in French Polynesia. It is the site of some amazing work by scientists on how species form by natural selection. The research has lasted so long that it shows some of the ways in which science has changed. It also shows how natural selection cannot always save organisms when conditions change very fast.

Moorea is one of the Society Islands, which include Tahiti

Studying *Partula*

Partula (Polynesian tree snails) have been studied on Moorea by the same scientists since the 1960s. When the research began 58 species of *Partula* on the islands had evolved to fill the different potential habitats. Scientists wanted to see the point at which natural selection produced a new species. The geographical isolation of Moorea and the number of snails it had made it an ideal place in which to study this.

Brian Clarke and James Murray identified the species of snails by looking at particular proteins. When they started their research it took weeks to analyse a single protein. They wrote their results down and posted them back to the UK.

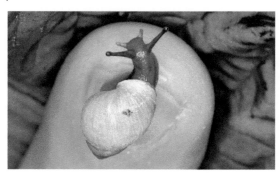

One type of *Partula* (Polynesian tree snail)

The first problem for the scientists was that it soon became possible to analyse not just proteins but the genetic material DNA itself. DNA analysis can now be done very quickly. All of the protein evidence the scientists had taken years to collect was suddenly out of date, as the scientific community required DNA evidence. Clarke and Murray revisited their snails taking DNA samples. Results could be sent by email in seconds.

However, a much greater problem had occurred. The giant African land snail was introduced to one of the local islands as a possible food animal. Unfortunately it escaped and spread to Moorea. It is a non-native species but it didn't compete with *Partula*. However it began destroying the crops and the garden plants all over the island.

The giant African land snail didn't affect *Partula* but it rapidly became a pest to the human population of Moorea and the other Polynesian islands

The real disaster came when people introduced another non-native snail to the island in 1977. This was a carnivorous Rosy Wolf Snail. The idea was that it would attack the giant African snails and control them. *Euglandina* was much smaller than the Giant African land snails – but much bigger than *Partula*. It made no attempt to attack the giant snails, but immediately started to destroy the small local tree snails and killed them rapidly. In three years it had cleared nearly a third of the island of snails.

By 1986, all of the local island species of *Partula* had gone forever. There are only five species left in the wild – and they are disappearing fast. Brian Clarke and James Murray managed to rescue 15 species of snails from the islands. For these species, the only specimens left alive are in breeding colonies in zoos around the world. The chance to see species forming has been lost.

The challenge for scientists now is to keep the colonies alive until the snails can be taken back to their natural habitat. That will only be possible if and when *Euglandina* finally becomes extinct on the islands.

The Rosy Wolf Snail (*Euglandina*) is a fierce predator

Summary questions

1 Suggest reasons why there are so many unique species of snails on Moorea and the Society Islands.

2 Give three ways in which scientific research changed during the years that Clarke and Murray were studying the snails on Moorea.

3 Make a poster to show the story of the snails on Moorea.

1 Look carefully at the photographs of the four members of the cat family that are found in the Masai Mara. They are on page 46.

a Using only the information on page 47, write a branching key to identify these four species. [4]

b Write out your branching key as a series of questions in a table, like that shown on page 49. [4]

c Lions feed on wildebeest and antelopes. State the trophic level of each of these three animals. [3]

d Explain how the four species of cat described on page 47 are able to survive in the Masai Mara even though they are very similar in appearance and feeding habits. [3]

2 a The cells of the house mouse *Mus musculus* each have 40 chromosomes and about 25 000 genes.

i Name the part of the cell that contains chromosomes. [1]

ii Name the chemical substance that makes up genes and chromosomes. [1]

iii State the role of genes in an organism, such as a mouse. [1]

iv Explain briefly how characteristics are passed from parents to their offspring. [3]

Oldfield mice, *Peromyscus polionotus*, live in the south-east region of the USA. They are brown in colour. Beach mice are the same species, but are much lighter in colour being yellow or white. Beach mice live on sand dunes and islands along the coasts of Alabama and Florida in the USA.

b Explain why the beach mice that live on sand dunes are much lighter in colour than the oldfield mice. [2]

c Scientists think that some oldfield mice migrated onto the sand dunes and over many generations the fur became much lighter to be like that of the beach mice today.

Suggest how this change in fur colour may have occurred. [4]

3 The following passage describes some of the feeding relationships in the Ruwenzori Cloud Forest in Uganda.

Green mountain bamboo is a plant that grows on steep slopes and is eaten by speckled cockroaches, okapi and colobus monkeys. Okapi look like antelopes but are closely related to giraffes. Ruwenzori turacos are brightly coloured birds that feed on the fruit of many forest shrubs, trees and lianas. Martial eagles eat colobus monkeys, turacos, honey badgers and young okapis. Honey badgers eat cockroaches and black mamba snakes. Black mamba snakes eat colobus moneys and young turacos.

a Draw a food web to show all the feeding relationships described in the passage. [5]

b How many food chains are there in the food web you have drawn? [1]

c From the food web identify one example of each of the following:

i producer, ii primary consumer, iii secondary consumer, iv tertiary consumer, v herbivore, vi carnivore. [6]

d Suggest what might happen to the food web if colobus monkeys became extinct in the Ruwenzori Forest. [3]

4 Rashmi and Kasun are part of their school's gardening club. They have discovered that allowing dead leaves to decompose will provide a good source of nutrients for the plants that they want to grow. They decide to put all the dead leaves from the garden into a compost bin, which is shown in the diagram.

a Name two groups of organisms that decompose the leaves in the compost bin. [2]

b Explain why the compost bin has holes in it. [2]

c Explain how dead leaves are a good source of nutrients for the school garden. [3]

5 a The photograph shows a Barbados blackbelly sheep. This breed was developed by farmers on Barbados from sheep introduced from Africa and Europe in the seventeenth century. The farmers used selective breeding to produce animals that gave plenty of meat and were adapted to the hot climate by not having thick wool.

a Explain what is meant by *adaptation*. [2]

b It is thought that farmers in Barbados used African sheep and woolly European sheep to breed the blackbelly. Suggest how they might have done this. [5]

6 During his voyage around the world in HMS Beagle, Charles Darwin noticed that the species he saw on islands were often different from similar species that he saw on the mainland. The bird in this photograph is a mockingbird from one of the Galápagos Islands. Darwin noticed that birds like this differed from island to island. He thought that mockingbirds of one species had migrated from South America and natural selection was responsible for the four different species that live on these islands.

Explain how natural selection could be responsible for the four different species of mockingbirds on the Galàpagos Islands. [6]

Science *in context!*

Population problems

Sometimes populations grow very quickly. Sometimes the numbers fall. Whatever happens, changes in populations can cause problems.

Too many insects ...

When locust populations grow too big, the animals form a swarm. Locust swarms can be miles long, containing millions of insects. The huge population of locusts eats all of the plant material in its way. Even a small swarm of locusts can eat as much food in a day as 35 000 people.

In 2004, a large proportion of the crops of the island nation Mauritius were destroyed by a massive locust swarm

... and too few

All over the world the population of honey bees has fallen. No-one knows exactly why this has happened. Scientists think that it is partly due to the use of pesticides and partly due to a new and mysterious disease that kills off whole hives of bees. This fall in the bee population is serious. Bees make honey, which people enjoy eating. But much more importantly, bees pollinate flowers, including most of the fruits we like to eat. Without bees there would be no pollination – and no fruit.

In China the bee population has fallen so low that farmers have to pay people to pollinate their apple and pear trees to make sure they get a crop

The rise and fall of a population

In 1944, US coastguards took 29 reindeer to St Matthew Island off the coast of Alaska. They were to act as extra food for the men stationed on the small island. The men left the island at the end of the Second World War. The reindeer stayed, eating the lichens that grew all over the island. By 1957 there were 1300 reindeer. Six years later, in 1963, the population reached 6000 – but there was hardly any lichen left. As the reindeer population grew, the lichen population fell. When scientists returned in 1966, there were only 42 reindeer left. There was only one male, who was not healthy enough to reproduce. Within a few years all the reindeer were dead.

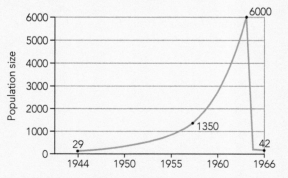

The reindeer population of St Matthew Island, 1944–66

In this chapter you will find out about the growth of populations, including the human population. You will discover some of the factors affecting population growth. You will also find out about some of the problems that arise when a population gets too big.

Key points

- A population is a group of animals or plants of the same species living in the same habitat.
- The numbers of organisms in a population will vary as a result of factors in their environment e.g. temperature, light or water.
- The amount of food available, competition for resources, disease, predation and overcrowding can all have a big effect on population numbers.
- Different populations will be linked in a community e.g. numbers of predators and their prey.
- The human population has grown enormously in the last two centuries.
- The global population explosion is the result of people being able to grow more food, being able to cure or prevent many diseases and having no natural predators.
- The increase in the human population has had a number of effects on the environment as a result of overcrowding.
- People are using much more land for housing, farming, etc. and this affects the organisms which used to live there.
- People are destroying forests to get more land for farming and homes – this releases carbon dioxide into the atmosphere and reduces the amount of carbon dioxide that is removed from the atmosphere by photosynthesis. Carbon dioxide is a greenhouse gas.
- The extra greenhouse gases may be causing global warming.
- Deforestation is leading to loss of biodiversity.
- The increase in the human population means more demand for water.
- Clean drinking water is needed to reduce the spread of disease.
- Pollution of water comes from human wastes, factories and farming (fertilisers and pesticides).
- Clean water can be provided by boiling, adding chlorine or bleach, digging wells used only for drinking water and piping water to homes.
- As more people use water, conservation of water and reusing water becomes more important.

Most of us think hedgehogs are rather appealing animals. But what most of us don't know is that hedgehogs are the habitat of the hedgehog flea! How many fleas do you think a single hedgehog will support?

The long-eared hedgehog is cute, eats insects – and carries fleas

Biological populations

The fleas on a hedgehog are called a **population**. This is a group of animals or plants of the same species living in the same **habitat**. If we can find out how many organisms are in a population, we can then see if their numbers go up or down. This in turn tells us something about the conditions in the environment. A healthy hedgehog will carry a lot of fleas!

The numbers of animals and plants in a population will vary as a result of all sorts of changes in their physical surroundings (**environment**).

Temperature has a big effect on populations. Most animals and plants grow faster when it is warmer so tadpoles like this turn into frogs faster in warm countries than they do in cooler ones. The sooner they become adult frogs, the more time they have to feed and grow and the safer they are from predators. This means more of them will survive, and the frog population will go up.

All sorts of factors will affect how quickly this tadpole turns into a frog

Light also affects the numbers of organisms, particularly plants. If there is less sunlight than usual the growth of plants is slowed and they are less likely to flower and produce seeds, which reduces the size of the future population.

Environmental damage after an enormous tsunami

Look at the photogragh at the bottom of page 70 and answer the following questions:

- What do you think caused the loss of the tree population in this photograph?
- What other populations might have been affected at the same time?

The amount of water available can make a huge difference to the populations of animals and plants in a habitat. We can see this in the mass of flowers that grow in the desert after a rare downfall of rain. The water allows thousands of seeds to germinate and grow into plants, which not only flower and make more seeds, but also feed many animals. The animals can then successfully raise their young and increase the size of their population too.

Population changes

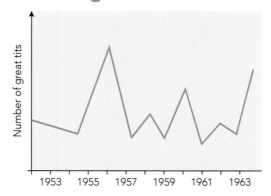

Graph to show the changes in the Great Tit population in Sweden over a 10-year period

Great Tit

As we can see from this graph, the numbers in a population don't stay the same every year. As the environment varies, in some years more great tits survive than in others.

It is not only the physical conditions of the environment which affect the populations of living things. The activities of other animals and plants – and of people too – can have a big effect on populations. The next pages will show you how.

Key terms

- **environment**
- **habitat**
- **population**

Summary questions

1 a) What is meant by a population?
  b) Give three examples of populations of animals or plants.

2 Look at the graph above showing changes in the Great Tit population in Sweden over ten years and use it to answer these questions:
  a) In which year do you think conditions were particularly good for the Great Tits?
  b) In which year do you think they were really bad?
  c) Suggest some environmental factors which might have affected these small birds which eat insects, seeds and berries.

In any population of animals or plants the numbers will change over time. The population will go up if lots of babies are born, and will go down if lots of individuals die.

Population growth or fall?

There are a number of factors which affect the balance of the population. They include:

- **The amount of food available.** If there is plenty of food for all (or light and minerals for plants) the population will increase. If there is not enough food, the organisms will not reproduce and may even die, so the numbers fall.
- **Competition for resources.** If there is competition between two different species for the same resources, the numbers of both species will stay relatively low. If there is no competition and lots of resources the population will grow.
- **Disease** can have a big effect on populations. A new disease can wipe out most of a population.

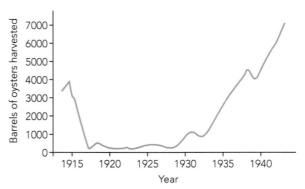

When a new Malpeque disease infected the Canadian oyster beds it almost completely wiped them out. Natural selection meant a new population of oysters resistant to the disease eventually grew up.

- **Predation** affects a population, particularly if a new and successful predator comes into an area. The numbers of the prey animals can fall dramatically (see topic 3.9 on the Rosy Wolf snail and the effect it has had on the population of *Partula* snails).
- **Overcrowding** happens if a population grows too quickly. There are a number of effects, all of which tend to make the population fall again. There may be a shortage of food, diseases spread easily between an overcrowded population and levels of waste may build up and poison some of the remaining population.

A community effect

All populations in a habitat interact with each other. What happens to one population in a **community** will always affect another. You see this very clearly in the ways that the populations of a predator and its prey are linked together.

Brown tree snakes were spread to the island of Guam by mistake. They have driven several species of birds to extinction and the population of other species is very low – all eaten by the snakes. The snake population is now very big indeed.

Think about the relationship between antelope and lions. If there is plenty of plant food the numbers of antelope (the prey animal) go up. This means there is more food for the lions (the predators), so more of their offspring survive and their numbers go up. But the extra numbers of prey overgraze the grass so there is less food and the extra numbers of predators eat more prey. As a result the antelope population falls. This in turn means there is less food for the lions, so their numbers fall. But with fewer antelope the grass recovers. With fewer lions and more food, the antelope population starts to grow again. This cycle tends to be repeated over the years.

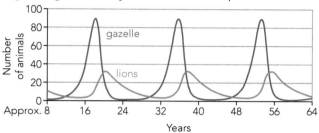

The population numbers of lions and antelope are closely linked

Practical activity   Looking at data

Look at the graph below and explain what is shown by this data.
- Why are there more rabbits than coyotes at first?
- Suggest reasons for the changes in the two populations.
- Sketch a graph which shows what you would expect to happen in the next few years.
- Use secondary sources to find out about how rabbits and coyotes interact. How reliable do you think this data is? What more would you want to know?

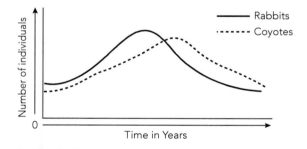

Key terms

- **community**

Summary questions

1 **a)** How does the availability of food affect a population?
  b) Give two other factors which can also affect the numbers in a population of organisms.

2 Look at the graph opposite to show the effect of Malpeque disease on oyster numbers.
  a) When did Malpeque disease affect the population?
  b) Suggest a reason why the oyster population became so small.
  c) How did the population recover?

3 Using secondary sources, investigate the relationship between the populations of snowshoe hares and lynxes in Canada. Explain how the data was collected and draw a graph to show the way the populations are linked. Make notes to explain what is happening at each stage of the graph.

4.3 Human population growth

Learning outcomes

After this topic you should be able to:

- describe the growth of the global population over time
- explain the reasons for the rate of increase and some of the disadvantages of this population growth.

For many thousands of years there weren't many people on the Earth. The human population was quite steady and there were only a few hundred million of us. People were scattered about all over the world. Whatever they did had only a very small and local effect.

A population explosion

In the last 200 years or so, the numbers of people on the Earth has grown very quickly. By 2012, the global population of the Earth was over 7 billion people – and it was still growing.

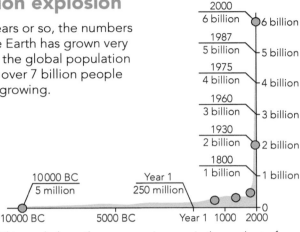

This graph shows the enormous increase in the numbers of people alive over the last few centuries

If the population of any other species of plants or animals suddenly grew like this, other factors would reduce it. Predators, lack of food, build-up of waste products or diseases would kill lots of people and the population would fall again.

However this does not happen with the human population. Why has the human population grown so fast and so much?

- We have discovered how to grow far more food than we could ever gather from the wild, so over much of the world people have food to eat.
- We can cure many diseases using **antibiotics** and other medicines.
- We can prevent many killer diseases using **vaccinations**.
- People have no natural predators.

This all helps to explain why the human population has grown so fast.

 IGCSE Link...
You will learn more about the effects of the increasing numbers of humans upon the natural environment.

The global population is very big – and getting bigger every day

The effect of so many people

There are so many people in the global population that now we have a big effect on our environment. Our activities seem to have changed the balance of nature on the whole planet. Some of the changes we have made seem to be driving other species to become extinct. There are some big disadvantages to having such a large population. These include:

- All of the seven billion people need land to live on. There is a limited amount of land on the Earth. People are taking land to build homes, shops, schools, factories and roads. This takes away the habitat of other organisms.

- Around the world billions of acres of land are used for growing food. The natural animal and plant populations are destroyed to make way for huge populations of crop plants.

- People quarry and mine to get many of the resources of the Earth, from oil to metal ores. This also has a big effect on the populations of other species, destroying their habitats.

In new supercities like Guangzhou, in southern China, millions of people live close together, but it still takes a great deal of land

---

Practical activity   Find out more

Use secondary sources such as books and the internet to find out more about the growth of the human population and the ways it is affecting the planet.

- Make a poster to explain EITHER how the human population has grown and changed over time OR how the human population is affecting the ecology of the Earth.

Key terms

- **antibiotic**
- **vaccination**

Summary questions

1 Suggest reasons why the human population has grown so rapidly over the last two hundred years when it remained stable for thousands of years.

2 a) Using secondary sources such as books and the internet, find predictions for the global population in the future.

   b) Do they all suggest the same figure?

   c) Suggest reasons why it is difficult to predict what will happen to the global population in the future.

3 Give three disadvantages for other organisms caused by the increase in the human global population.

After this topic you should be able to:

- explain why deforestation is a growing problem
- describe the effects of deforestation
- describe the effect of habitat destruction on other species.

As the global population grows, people need more land, more food and more fuel for their cars. To help solve this problem, huge areas of forests, particularly **tropical rainforests**, are destroyed by **deforestation**.

The effects of deforestation

When people want to use the land for farming, the trees are often just cut down and burned ('slash-and-burn' clearance). The wood isn't used for anything. The land produced isn't fertile for very long. No trees are planted to replace those cut down. The land is often used to grow extra palm oil or beef for people in developed countries rather than food crops for the growing populations where there is not enough food to go around.

Deforestation also increases the amount of carbon dioxide released into the atmosphere. If the trees are burned, this makes carbon dioxide. If the trees are left to rot, the decomposers make carbon dioxide.

Cutting down the trees also means that they can't use carbon dioxide for photosynthesis. Normally, large amounts of carbon dioxide are locked up in the biomass of the trees, sometimes for hundreds of years. So when we destroy trees, we destroy a way of taking carbon dioxide out of the air. If the land that is cleared is used to farm cattle, the cattle add even more to the carbon dioxide in the air.

Gases such as carbon dioxide trap a layer of warm air around the Earth, which is vital for all life on Earth. This is known as the **greenhouse effect** and carbon dioxide is a **greenhouse gas**. However the evidence shows that the concentration of carbon dioxide in the atmosphere is increasing all the time. This is causing an enhanced greenhouse effect because the extra carbon dioxide we are adding is trapping more heat. This may be the cause of **global warming**. Scientists think global warming causes changes in the climate, extreme weather and the melting of the polar ice caps. The burning of fossil fuels such as gasoline in cars also produces a lot of carbon dioxide.

'Slash-and-burn' clearance

Deforestation

Atmospheric Carbon Dioxide
Measured at Mauna Loa, Hawaii

The level of carbon dioxide in the atmosphere is increasing every year

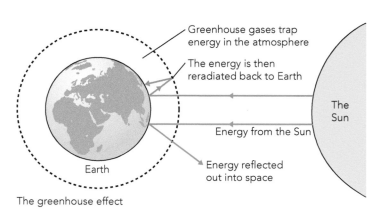

The greenhouse effect

Make a poster on the problems of global warming linked to deforestation. Include as much data as you can. Look carefully at the different evidence you can find in secondary sources such as the internet. How will you decide which of the data is reliable?

Habitat loss

When we cut down forests we destroy the habitat of all the animals and plants that lived in them. A single tree can be the home of hundreds of different types of organisms. Tropical rainforests are especially important because they contain more **biodiversity** (variety of living organisms) than any other land habitat. If we destroy the habitat of an organism it may become extinct. This is already happening in many countries.

Solving the problems

It is not easy to stop deforestation. Populations in the developed world can stop demanding the products grown in the reclaimed land. Countries with rainforests can refuse to have them cut down, but many of these countries are poor with fast growing populations and they need the money.

People in the developed world could pay to preserve the rainforest. As local people become more aware of the importance of their forests, perhaps they will protect them. Replanting local species of trees can help. Education and awareness are important all over the world.

Scientists think extreme weather caused by global warming may cause problems such as this flooding in Pakistan

Practical activity   Debating the issues

Work in a small group. Using secondary sources such as books and the internet find out as much as you can about deforestation and possible ways to overcome the problems it causes. Then write a report for the television news.

Key terms

- **biodiversity**
- **deforestation**
- **global warming**
- **greenhouse effect**
- **greenhouse gas**
- **tropical rainforest**

Summary questions

1 a) What is deforestation?
  b) Suggest why the increase in human population led to an increase in deforestation.

2 Explain the possible links between deforestation and global warming.

3 What is biodiversity and how is it affected by deforestation?

People need water to drink. The more people there are, the more water is needed. The water that people drink must be clean. About 2.2 million people die of sickness and diarrhoea every year, mostly children under five years old and mainly as a result of drinking water carrying disease.

This water may well be full of the microbes that cause sickness and diarrhoea

Water pollution

The more people there are in the population, the more problems there are with water **pollution**.

Sewage (human waste) is often released into the water in rivers or seas. Human waste carries many diseases. If people drink water contaminated with sewage they can quickly become ill. Sewage also makes the water high in nitrates and phosphates. This can lead to a lack of oxygen and the loss of all the animals in the water.

Practical activity   Find out about eutrophication

Using secondary sources, find out about the process of eutrophication, which can be caused by too many nitrates in the water.

Make a poster to display your findings.

Eutrophication – eventually there will be no animal life in the water at all so the river or pond becomes dead

- Factory wastes may be released into waterways and get into the water that people drink. This can cause serious health problems and even death.
- **Fertilisers** and **pesticides** used by farmers to grow as much food as possible can easily wash out of the soil into the rivers and lakes which we use for our drinking water. Fertilisers contain nitrates and phosphates and can cause the same problems as sewage. Pesticides can get into food chains and cause disease in the top predators.

Solving the problem of water-borne disease

The bigger the human population in an area, the more clean water is needed and the more sewage is produced. Sickness and diarrhoea from dirty water kills more children than any other disease.

If people understand that disease is spread by drinking water containing microorganisms from body wastes, then they are more likely to try one of the following solutions:

- **Making drinking water safe.** Treating the water with chlorine before it goes into pipes will kill off any microbes. Simple wells can be dug to allow people to get clean water close to their own homes. Boiling water or adding a small amount of bleach can make the water safe to drink.

- **Improved sanitation**. The sewage treatment programme can be improved so human waste is treated and made safe before it returns to the water supply. Pit latrines for body waste are a good way of reducing the spread of disease if they are placed well away from any drinking water source.

If wells are sited properly and people use them ONLY for drinking water, they can make a very big difference to the spread of disease

Conserving water

Water is a limited resource – the amount of rain that falls does not increase as the human population goes up. What is more, around the world more and more people are using modern appliances such as flush toilets, baths, showers and washing machines. All of these use water. People are looking at ways to conserve the water we have so that everyone can have clean drinking water as well as modern appliances. Ways of conserving water include reusing water from washing to flush toilets, collecting rainwater for washing and flushing toilets, and using showers which use less water.

> **Practical activity**   How can we save water?
>
> Work in a group to produce a booklet which could be given to all the households in your community to help them conserve water.
>
> Explain the importance of conserving water and suggest as many ways as possible in which people can save this important resource.

> **Key terms**
>
> - **fertiliser**
> - **pesticide**
> - **pollution**
> - **sewage**

> **Summary questions**
>
> 1 How has an increase in the human population affected the water supply?
>
> 2 **a)** Why is it so important to have clean drinking water?
>   **b)** Suggest two ways to make sure that drinking water is clean and safe to drink.
>
> 3 What is water conservation and why is it becoming more important?

1 Match the following terms with their definitions:

A population
B environment
C habitat
D deforestation
E competition
F community

1 all the features that make up an organism's surroundings
2 the place where an organism lives
3 all the organisms that live in the same place at the same time
4 all the organisms of the same species that live in the same place at the same time
5 the removal of trees from an area
6 the struggle between organisms for resources [6]

2 A university student was studying the numbers of kangaroos and dingoes that visited a watering hole in a National Park in Australia. Kangaroos are herbivores and dingoes are carnivores that hunt kangaroos as well as eat other prey.

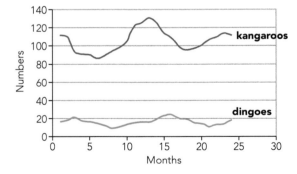

a Describe the changes in the numbers of kangaroos and dingoes. [4]
b Do you think the changes are similar to those of predators and prey as described on page 73? Explain your answer. [3]
c Evaluate the way in which the student collected the data on the two species. [4]

3 The table shows the estimated human population between the years 1000 and 2012.

Year	Estimated human population / billions	Year	Estimated human population / billions
1000	0.3	1960	3.0
1500	0.4	1975	4.0
1600	0.5	1985	5.0
1700	0.6	2000	6.0
1800	1.0	2012	7.0
1925	2.0		

a Draw a graph to show the change in the human population. [5]
b Give three reasons why the rate of increase in the human population was much slower in the years before 1800 than afterwards. [3]
c Many advances in medicine, such as the development of antibiotics, were introduced after 1940. Explain why the human population began to increase long before that. [3]
d State four resources that humans need from the environment that are put under pressure as human numbers increase. [4]

4 A small island country imports its energy in the form of oil and coal that are burnt in power stations to provide electricity for the population. The island has steep mountains, high rainfall, fertile soils with good growing conditions for crops and strong offshore winds.

a Explain why coal and oil are called fossil fuels. [2]
b Describe the effects of burning coal and oil on the environment. [3]
c The government has decided to make use of the country's resources to meet its increasing energy needs.
  i Explain why you think it has made this decision rather than import more oil and coal. [3]
  ii Suggest how the government can make use of the island's resources to meet the country's energy needs. [4]

iii Many people on the island object to
these developments. Suggest the
objections that they might have.   [3]

5 The diagram shows a river that flows through
agricultural land and past a settlement from
where it receives untreated sewage. Further
downstream water from the river is taken
and treated to provide drinking water for a
nearby town.

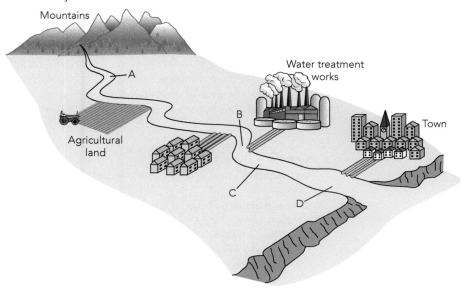

Scientists from the environmental protection
agency took samples in the river. They
counted the numbers of some of the animals
that they found. Their results are in the table.

Sampling site	Stonefly nymphs	Midge larvae	Rat-tailed maggots	Minnow (small fish)	Sludge worms
A	17	3	1	6	3
B	0	38	9	0	78
C	0	36	5	1	39
D	7	14	0	3	10

a Which animals were found in all of the
samples?   [2]

b How does the sample taken at B differ
from the sample at A?   [3]

c The scientists found that there were huge
numbers of bacteria and fungi in the water
at points B and C, but few at A and D.
Explain why.   [3]

d Which of the animals are best adapted
for living in polluted waters? Explain your
answer.   [3]

e The farmer decides to use chemical
fertilisers on his land between A and
B. The scientists think that much of the
fertiliser will run off into the river. Suggest
the effects that this might have on the
river.   [3]

f The scientists think that it would be a
good idea to move the water treatment
plant upstream to A. What are the
advantages and disadvantages of this
idea?   [3]

Glossary

A

Adaptation Special feature that makes an organism particularly well suited to where it lives.

Anther The part of the plant that makes the pollen containing the male gametes.

Antibiotic A drug that destroys bacteria inside the human body without damaging human cells and so cures many bacterial diseases.

Antibiotic resistance Bacteria which have become resistant to an antibiotic and are no longer killed by it as a result of natural selection.

B

Biodiversity A measure of the number and variety of different organisms found in a specific area.

Biomass The amount of biological material in an organism or a trophic level.

C

Carnivore An animal that feeds on other animals.

Carpel The female sex organs in plants.

Charles Darwin The scientist who first developed the theory of evolution by natural selection.

Chlorophyll The green pigment (colour) found in plants that traps light energy to be used in photosynthesis.

Chloroplast The organelle that contains the chlorophyll in the leaves.

Chromosome A structure found in the nucleus of the cell made of DNA containing thousands of genes.

Community All the organisms of different species living together in the same place.

Cross-pollination The transfer of pollen from the anther of a flower to the stigma, or a flower on another plant.

D

Decomposer Organism that feeds on dead bodies and waste and uses it as food.

Deforestation The cutting down of large areas of forest without any replanting.

Destarch The process of removing the starch from a plant before carrying out photosynthesis experiments by leaving the plant in the dark for about seven days.

Dispersal The process by which seeds are spread as far as possible from the parent plant.

DNA The chemical that makes up the genes.

E

Embryo plant The tiny new plant found in a seed.

Endothermic reaction A reaction in which more energy is taken in than given out.

Environment The physical surroundings an organism lives in.

Epidermis The upper protective layer of cells in a leaf.

Exothermic reaction A reaction in which more energy is given out than taken in.

F

Fertilisation The fusing of male and female gametes to make a new individual, e.g. in plants the fusing of the nucleus of the pollen grain and the ovule.

Fertiliser Substance which is used to add mineral salts to the soil and make it more fertile.

Filament The stalk that attaches the anther to the flower.

Food web A number of food chains linked together to show the feeding relationships between the organisms.

G

Gene The unit of inheritance.

Germination When the new plant begins to emerge from a seed.

Global warming The slow rise in the average temperature at the surface of the Earth as the level of greenhouse gases in the atmosphere increases.

Greenhouse effect The warming effect of gases such as carbon dioxide and methane in the atmosphere. They act like a greenhouse to retain warmth at the surface of the Earth so life can exist.

Greenhouse gas A gas found in the atmosphere that traps heat close to the surface of the Earth, e.g. carbon dioxide, methane.

H

Habitat The place where an animal or plant lives.

Herbivore An animal which feeds on plants.

Hydroponics Method of growing plants in mineral enriched water without any soil.

I

Iodine solution A red-brown liquid, which changes to dark blue/black in the presence of starch.

Irrigation Bringing water to soil artificially through ditches or pipes and pumps.

K

Key A tool you can use to help you identify and classify organisms.

L

Limiting factor A factor that can restrict how fast a process, such as photosynthesis, can work.

N

Natural selection The process by which organisms best adapted to their environment survive and breed to pass on their characteristics to their offspring.

Nectar The sweet sugary liquid produced by some insect pollinated flowers to attract insects.

Nitrogen Main gas in the air. Also a key element for plant growth in the form of nitrates.

O

Ovary The part of the carpel that produces ovules.

Ovule The female gamete in plants.

P

Parasite An organism that feeds off another living organism, causing it harm.

Peppered moth A moth which shows clear evidence of natural selection between the pale and the dark forms.

Pesticide A chemical that kills pests.

Petal Structures which protect the sex organs of a plant and may be brightly coloured to attract insects.

Phosphorus A key element needed for plant growth in the form of phosphates.

Pollen grain Structure which contains the male gametes in plants and fuses with the female gamete in the ovule.

Pollen tube The tube which grows out of a pollen grain down the style and into the ovary so the pollen nucleus can travel down it and fuse with the ovule nucleus.

Pollination The transfer of pollen from the male parts of one flower to the female parts, usually of another flower.

Pollution The contamination of the natural environment by harmful substances as a result of human activity.

Population A group of organisms of the same species living in the same habitat.

Potassium A key element in plant growth, taken into plants in the form of potassium salts.

Primary consumer An animal that eats plants.

Producer An organism that produces its own food by photosynthesis.

Pyramid of numbers A way of representing the numbers of organisms at different stages of a food chain.

R

Reliable Data that is repeatable and can be reproduced.

Repeatable An experiment that can be repeated by the same experimenter and give the same results.

Reproducible An experiment that can be carried out by a different experimenter and they get the same results as the original experiment.

S

Secondary consumer An animal that eats herbivores.

Selective breeding Choosing individuals with characteristics you want to be the parents of the next generation, to improve a plant or animal.

Self-pollination The transfer of pollen from the anther of a flower to the stigma of the same flower, or to a stigma of a flower on the same plant.

Sepal Structures that protect the flower bud, usually green.

Sewage Human bodily waste and waste water from homes.

Stamen The male sex organ in plants.

Stigma The area in the female part of the flower where the pollen grains land in pollination.

Stomata (single stoma) The openings in the surface of a leaf which allow gases to diffuse in and out of the leaf.

Style The stalk which attaches the stigma to the ovary.

T

Tertiary consumer A carnivore that eats other carnivores.

Tropical rainforest A type of forest found in tropical areas that contains a very large number of species of animals and plants.

V

Vaccination The process of introducing small amounts of dead or inactive pathogens in to the body to produce immunity to future infections.

Valid An experiment which is suitable to answer the question it is investigating.

Variable Physical, chemical or biological characteristics which can be changed during an investigation.

Index